AF712161

Michael Has

Sustainable Products

Also of Interest

Energy and Sustainable Development
Quinta Nwanosike Warren, 2021
ISBN 978-1-5015-1973-4, e-ISBN 978-1-5015-1977-2

Machine Learning for Sustainable Development
Kamal Kant Hiran, Deepak Khazanchi, Ajay Kumar Vyas
and Sanjeevikumar Padmanaban (Eds.), 2021
ISBN 978-3-11-070248-4, e-ISBN 978-3-11-070251-4

Sustainable Process Engineering
Gyorgy Szekely, 2021
ISBN 978-3-11-071712-9, e-ISBN 978-3-11-071713-6

Blockchain 3.0 for Sustainable Development
Deepak Khazanchi, Ajay Kumar Vyas, Kamal Kant Hiran
and Sanjeevikumar Padmanaban (Eds.), 2021
ISBN 978-3-11-070245-3, e-ISBN 978-3-11-070250-7

Michael Has

Sustainable Products

Life Cycle Assessment, Risk Management,
Supply Chains, Eco-Design

DE GRUYTER

Author
Distinguished Professor
Maitre de Recherches
Michael Has
Michael.Has@Protonmail.com

ISBN 978-3-11-076729-2
e-ISBN (PDF) 978-3-11-076730-8
e-ISBN (EPUB) 978-3-11-076739-1

Library of Congress Control Number: 2021948041

Bibliographic information published by the Deutsche Nationalbibliothek
The Deutsche Nationalbibliothek lists this publication in the Deutsche Nationalbibliografie; detailed bibliographic data are available on the Internet at http://dnb.dnb.de.

Cover image: Gettyimages / Nannapat Pagtong
Typesetting: Integra Software Services Pvt. Ltd.
Printing and Binding: LSC Communications, United States

www.degruyter.com

Preface

This book is based on lectures on circular economy, life cycle analysis and determination of CO_2 – and material footprints as well as eco-design given at INP Grenoble at the Department of Paper and Biomaterials between 2015 and 2020.

The industry's recent orientation towards ecologically sustainable production without neglecting economic necessities seems comparatively new, but goes back to a longer development. Therefore, the following text begins with a brief historical overview in which social, economic and ecological developments are placed in a context. The approach taken here leans heavily on Fridell's considerations, according to which social, economic and technical developments must always be seen in context. An understanding of the background is essential to understand the direction of current public and political opinion and the resulting actions.

The second chapter aims at understanding basic concepts describing the flow of products in a linear and closed-loop economy. It also describes the eco-design of products as an enabler for a closed economy.

Chapter 3 focusses on different concepts for ecologically oriented business models.

Chapter 4 introduces life cycle assessments and material and energy accounting along the phases within life cycles. Balancing, recording and calculation of environmental footprints (Chapter 5) corresponds to balancing consumption. The ultimate goal is to design products and processes in such a way that consumption is minimised. Concepts for this are often accompanied by more or less major changes in the design of production routes, products and their use. Priorities often have to be set, leading to the disadvantage of one parameter over the other. Chapter 5 provides examples of such assessments and Chapter 6 offers tools and an example of priorities in eco-design.

The previous chapters have only dealt with material products. Chapter 7 introduces the organisational environmental footprint that arises as a result of people working together in dedicated organisations. Of course, footprint reduction is necessarily closely tied to what is produced and the processes used – still some approaches are useful to keep in mind when looking for potentials. Chapter 8 offers general approaches to reduce material and energy footprints.

Internally, larger organisations have different teams and departments working together. The reduction of footprints poses new challenges for the cooperation and management of departments. Chapter 9 presents examples of issues that need to be discussed between teams in order to optimise internal cooperation. In the external cooperation with partners along the supply chains, strategies have to be implemented and the contributions to the reduction of material and energy have to be checked – audits serve this purpose. Chapter 10 deals with such audits.

In order for a company to be successful with a "greener" approach, it is important that accurate and rather than greenwashed messages are communicated from the beginning of the change process until well after implementation. Chapter 11

https://doi.org/10.1515/9783110767308-202

therefore addresses the communication aspect of the efforts going along with the reduction of CO_2 and material footprints.

Instead of a conclusion, Chapter 12 discusses various forecasts for climate development and accompanying circumstances until 2050 – given what is known today it does not and cannot present an optimistic forecast.

As indicated, this book is based on lectures held at INP Grenoble and, as always in teaching and writing, aside of students many persons contributed in various ways making the lectures and this book possible. In particular, it is a pleasure to thank

- Dr Bernard Pineaux for his ongoing support and encouragement for teaching and students.
- Dr Uwe Has, Jeannette Berger, Johannes Steigerwald and, especially, Martin Thompson for detailed cross reading and very valuable feedback – especially Martin corrected for misleading wording with sheer never-ending patience.
- Dr Barbara Heikel for drawing my attention on items I did not look at before, her willingness to share scientific and technical expertise and often challenging positions expressed in the manuscript.

Literature

I have provided many references that I hope readers will find useful. I hope I succeeded in giving a reasonable summary of the findings presented by the source. Should some of the ideas presented here be not referenced or repeat those of others without being cited, I apologise and can only ask that the relevant work be brought to my attention so that appropriate mention can be made in any subsequent editions of this book.

Contents

Preface —— V

Chapter 1
Historical development and basic ideas —— 1
References —— **15**

Chapter 2
Economic-ecological concepts —— 19
2.1 Value creation and supply chains —— **19**
2.1.1 Supply chain —— **19**
2.1.2 Technological progress in value chains – a dilemma —— **21**
2.2 Circular economy —— **21**
References —— **23**

Chapter 3
Business and sustainability —— 25
3.1 Business models —— **25**
3.2 Sustainable business models —— **25**
3.3 Schools offering theories on ecologically sustainable business models —— **27**
3.3.1 Natural capitalism —— **29**
3.3.2 Blue economy —— **30**
3.3.3 Regenerative design —— **31**
3.4 Criteria for evaluation and benchmarking for economic sustainability —— **32**
3.4.1 SDGs of the United Nations —— **32**
3.4.2 Measures and criteria for the transition phase —— **33**
3.4.3 Resolve methodology —— **34**
3.4.4 Emission pricing —— **34**
3.4.5 Assessment at national levels – beyond GNP —— **37**
References —— **39**

Chapter 4
Life cycle assessment (LCA) —— 43
4.1 Introduction —— **43**
4.2 Steps to take in the standardised LCA process —— **43**
References —— **48**

Chapter 5
Environmental footprints and their calculation — 51
5.1 CO_2 footprints — 51
5.1.1 GHG emission footprint and CO_2 footprint — 51
5.1.2 GHP Scope 1, 2 and 3 emissions — 52
5.1.3 Emissions — 54
5.2 Economic risks linked to emissions — 57
5.3 Example: print on different substrates — 66
5.3.1 Boundary conditions — 67
5.3.2 Printing — 73
5.3.3 Discussion — 80
5.4 Example: backing oven — 82
5.5 Example: digital newspaper — 86
5.5.1 Phases in lifetime — 86
5.5.2 Server farms — 88
5.5.3 Read/consume — 88
5.5.4 Discussion — 91
5.6 Summary — 91
References — 94

Chapter 6
Eco-design — 97
6.1 Guidelines — 97
6.2 Tool kits — 98
6.2.1 MET matrix — 98
6.2.2 Eco-design-strategy wheel (LiDS-Wheel) — 99
6.2.3 Eco-designchecklists — 101
6.2.4 An approach to eco-design — 104
6.3 Eco-design – a “high-level” example — 105
6.3.1 Introduction — 105
6.3.2 Approach — 106
6.3.3 Starting point — 106
6.3.4 Product life cycles and circular economy — 107
6.3.5 Recommendations on eco-design and management for the circular economy — 108
6.3.6 Eco-design guidelines — 111
6.3.7 Eco-management guidelines — 111
6.3.8 Recommendations for the management of environmental impacts — 111
6.4 Sustainable approaches to fibre-related resource management — 114
6.4.1 Order initiator supply chain checklist — 115
References — 117

Chapter 7
Life cycle assessment of an organisation — 119
7.1 Introduction — 119
7.2 General approach — 120
7.2.1 Priority principle — 120
7.3 Sustainability of an organisation — 121
7.3.1 Background — 122
7.3.2 Best practices — 124
7.3.3 Environmental audits for organisations — 125
References — 127

Chapter 8
Energy and material strategy — 131
8.1 Material strategy — 131
8.1.1 Accounting and return logistics — 131
8.1.2 Material recycling as a business model – an example — 137
8.2 Energy-related strategy — 140
References — 141

Chapter 9
Stakeholder and their contributions — 143
9.1 Manufacturing and logistics — 144
9.2 Research and development — 146
9.3 Facility management — 150
9.4 Procurement — 150
9.5 Marketing — 156
9.6 Quality management — 159
9.7 Health and safety — 162
9.8 Legal — 163
9.9 Distribution and logistics — 164
9.10 Field service — 166
9.11 Duties of all stakeholders — 167
References — 172

Chapter 10
Audits — 173
10.1 General — 173
10.2 Questionnaires — 176
10.2.1 General — 176
10.3 Reporting — 178
References — 179

Chapter 11
Information and transparency — 181
11.1 Communication with employees — 182
11.2 Communication with the public — 183
11.3 Communication with customers — 184
11.4 Communication with share holders — 188
11.5 Communication with partners — 190
11.6 Corporate sustainability report — 192
11.7 Environmental Policy — 192
References — 194

Chapter 12
Instead of a conclusion – a perspective? — 197
12.1 The Paris Agreement and the 2 °C target — 197
12.2 Tipping points — 199
12.3 Economic scenario until 2050 — 200
12.3.1 Fewer jobs and climate change lead to migration — 201
12.4 Legislation and litigation — 201
12.4.1 Legal disputes and court proceedings — 203
12.5 Summarizing thoughts — 204
References — 207

Annex

Annex 1 – Sources — 215

Annex 2 – Glossary — 217

List of Figures — 223

List of Tables — 225

Biography — 229

Index — 231

Chapter 1
Historical development and basic ideas

What can be observed in the present is based on roots, some of which go far back into history. Patterns and concepts are also often repeated. Approaches developed during the industrial revolution, for example, led to prosperity (at least for many in some regions of the world), often had long-lasting success and, like all success, came at a price. Finding a starting point for a historical discussion is subjective, regardless of which viewpoint one takes; furthermore, there is no ONE starting point. Without claiming to be exhaustive, this first chapter is intended to give an overview of relevant historical contexts. The following is about pointing out certain lines of development and relating them to their consequences to the modern world. It will therefore not proceed in the form of a strict historical sequence.

For this brief historical review, we will start with James Watts and his invention of the steam engine. His idea became one of the starting points of the industrial revolution, as it enabled the exploitation of natural resources in the context of industrial production. Arnold Toynbee used the term industrial revolution to describe the economic development of Britain from 1760 to 1840 (although the term industrial revolution was used earlier by French writers). It refers to the process of change from a partially circular agricultural and craft economy to a linear economy dominated by industry and machine manufacturing [37]. The term linear economy today relates to the way resources were harvested and wasted would be called a cradle-to-grave economy, today: Resources are harvested, used, consumed and subsequently become waste. The main features of the industrial revolution were technological, socio-economic and cultural. Technological changes were given in Table 1.1.

Table 1.1: Technological changes during the industrial revolution according to [37].

(1)	The use of new basic materials, mainly iron and steel – and more recently information
(2)	The use of new sources of energy, both fuel and motive power, such as coal, the steam engine, electricity, petroleum and the internal combustion engine
(3)	The invention of new machines that allows higher production with less human energy
(4)	A new organisation of labour, known as the factory system, involved an increasing division of labour and specialisation of functions
(5)	Important developments in transportation and communications, including the steam engine locomotive and steamship
(6)	The increasing application of science to industry

When speaking of the second, third or fourth industrial revolution, the intention is to refer to the renewed change in one or more of these parameters in an economic context. These changes enabled and caused a sudden increase in access to resources, their use and, in the context of specialisation the existence of an international

https://doi.org/10.1515/9783110767308-001

trade network, subsequently the mass production and consumption of goods. Expanded resource use, specialisation and mass production necessarily imply centralisation and this in turn implies a need for transport. Centralisation and transport characterise large economies to this day.

Political or economic power structures were initially unaffected by the technological and industrial development. During the time of the industrial revolution, in the third quarter of the eighteenth century, Adam Smith developed the idea of the "invisible hand" – referring to the self-regulating mechanisms in stable economic systems. In his 1776 work *An Inquiry into the Nature and Causes of the Wealth of Nations* [38], he used this idea again to describe the tendency of free markets to self-regulate. According to Smith, the influencing forces that determine market mechanisms are competition, supply, demand and self-interest. Ultimately, self-regulation is about the stability of the economic system to ensure that it delivers the maximum value possible from the point of view of the agents.

To this day, the idea of self-regulation as proposed by Smith is discussed very controversially: In particular, there is one aspect to pay attention to – Smith did not introduce nature as a limited resource as one of the forces involved in self-regulation. Taking today's point of view, this would have made sense, as such a view would have helped to draw attention to dangers emanating from an overexploitation of nature, that is, non-sustainable action.

These concepts were not new at the time: In this context, it should be pointed out that sustainability-like concepts were already known at the time of Smith, for example, the corresponding German term "Nachhaltigkeit" was used as early as 1713 by Hans Carl von Carlowitz (1645–1714) to describe the need to preserve natural forests for future generations in view of the shortage of the resource wood.

But the school of thought based on Smith's thinking did not take into account the environmental problems at hand. This was because the need for a stable nature/environment was not seen as a necessary part of the forces influencing the economy; the pre-Romantic period viewed nature merely as a backdrop, not as an economic asset. Although becoming less dominant, the mindset that a free market does not care about the environment can still be seen as more or less mainstream today. Times are changing, however – today it is undisputed that value can be ascribed to the stability of the natural environment. This thinking has prevailed in recent decades.

In this context, market regulation has also been accepted to address environmental concerns through conservative thinking.

The Reagan administration, certainly not to be seen as representing a left-wing movement, countered the desire to build new nuclear power plants with an argument that fits well with Adam Smith's theory:

If such a plant was safe, the thinking went, then it should be commercially insured. The risk to the environment associated with the operation of a nuclear power plant or nuclear waste repository should not be borne by the general public, but by the owner of the respective power plant. Consequently, the price of such insurance should be added to the price of the electricity produced – an argument which contributed to the reasons why the further expansion of nuclear energy in the USA was subsequently halted.

Although this can be seen as promising from the perspective of nature as a resource, there is no general trend here: In the USA, too, the risk of final destruction of land associated with uranium mining and initial processing was accepted – the so-called National Sacrifice Areas (zones) were declared to state that the respective area may no longer be used by humans. The idea that nature provides value and should therefore be cared for only became the subject of theoretical and political considerations later on (see Chapter 4).

Thomas Malthus published *An Essay on the Principle of Population* around 1800 [39]. He stated:

> I think I can justly state two postulates:
> First, that food is necessary to the existence of man.
> Secondly, that the passion between the sexes is necessary, and will remain almost in its present state.

From this Malthus derived two conclusions, namely,
(a) the food supply grows at an arithmetical (linear) rate, and
(b) population growth occurs at a geometric (exponential) rate.

In turn, he concluded that if the population grows faster than the required food supply, natural controls such as famines and wars will reduce the population to a sustainable level.

Malthus was publishing his insights at the brink of the industrial revolution in England. In the centuries before his lifetime, frequent and widespread famines were the norm. His theory was revolutionary because it explained conclusively why these famines had to occur. Malthus predicted that the next great famine would occur in the then coming decades. But it did not. The European population did not grow as Malthus predicted, and furthermore, improvements in agricultural technology (especially Liebig's observations that led to the development of fertilisers) allowed food supplies to grow more steadily than expected.

Malthus' thoughts were crucial for later thinkers such as Ricardo, Marx or Darwin; indeed, his theses continue to influence policy to this day. The population theory used by the Club of Rome around 1970 was partly based on a revised and expanded Malthusian model (revised by Michael Dewey's observation that crucial socio-technical developments such as hunting, the abandonment of nomadism, the development of agriculture, land reforms and industrialisation encouraged further population growth).

Another thinker to consider is David Ricardo with his book *On the Principles of Political Economy and Taxation* in 1817 [40]. In this publication, Ricardo developed the principle of comparative advantage in the context of the early theory of international trade. In this context, the basic assumption made in this publication is that the gains from trade follow from the fact that an economy can specialise. This means that in an example used by Ricardo, if a country is relatively better at producing wine (Portugal) than wool (Britain), it is rational to put more resources into wine and export some of the wine to pay for the import of wool.

It follows that in order to profit from trade, a country does not have to be the best. The gains come from specialising in those activities that the country can do better economically at world prices, even if it has no absolute advantage in them. It follows that trade barriers (taxes/tariffs) between these countries and in relation to these types of products should be reduced. This supports the view expressed above: transport is a necessary by-product of specialisation and concentration and an essential part of both the industrial revolution and internationalisation. To date, different countries have pursued very different strategies depending on whether they support such transport-related infrastructure (through subsidies or by building roads, etc.) and subsidies the use of fossil energy by offering tax advantages or whether they discourage it through taxation.

It is now considered proven that the number of countries participating in a free trade area should be as large as possible in order to stimulate all economies involved. Despite the fact that international trade contributes to welfare, it must be seen to do so at the cost of transport and extended alienation of consumers from the products they consume (in the sense that a consumer does not identify with the product consumed, see below), which in turn leads to consequences for ecology.

It is important to point out that this view on forces of influence identified by Adam Smith and the resulting need for centralisation were shared by Karl Marx and Friedrich Engels: What they sought was to change the underlying power structure. In the Communist Manifesto of 1848, Marx and Engels concluded that "the proletariat will use its political superiority to gradually wrest all capital from the bourgeoisie, to centralise all instruments of production in the hands of the state, i.e., of the proletariat organised as the ruling class; . . . in the course of development . . . the whole of production has been concentrated in the hands of a great association of the whole people" [41].

It is remarkable, but also a consequence of this way of thinking, that in countries that have practised state central planning of production, no increased responsibility for natural resources has become apparent (for the Soviet Union, e.g., Charles E. Ziegler [42]). One might think that since the state both "knows everything" and represents the people, it should be well placed to plan for environmental concerns. Historically, this has not happened, however.

The impact of industrialisation became very visible in the first decades of the nineteenth century. Before K. Marx looked at industrial mass production, specialisation and ownership of the means of production in his time, among other things, he observed alienation between production equipment and workers, which leads to careless handling of the resources worked/consumed due to the lack of identification between resources and production equipment in mass production. In the context of the strand of action pursued here, alienation between producers, consumers and what is produced is also worth mentioning (Christien Meindertsma [43]): while among workers and owners of production facilities the lack of emotional attachment leads to a certain carelessness towards the environment, the effect seen by Meindertsma relates to the identification between products and their consumers. This lack of relationship between products and their consumers leads to shorter product lifetimes and careless acquisition, use and scrapping of products (see also Chapter 4). All in all, alienation implies that all parties involved have different interests – which are not aligned for the mutual good or that of the environment.

With the rise of mass production, demand generation and differentiation became crucial ingredients of product development as new mechanisms to stimulate growth and bring more products into (partially saturated) markets. Although advertising, branding, packaging and labelling already existed in ancient times [45], the conscious application of market mechanisms identified did not develop until the turn of the century under the then new term *marketing*. It was not until after World War II that marketing was seen as an accepted means of generating broad public demand beyond existential needs [44].

Economic policy, if it existed at all, was not inherent in the basic principle of Adam Smith's theory of self-regulating markets initially. Most often, governments did not interfere in economic matters. Before World War I, governments primarily saw themselves as shields of private wealth. The German term "Nachtwächterstaat" describes this very well: an administration that concentrates on internal administration and external security to protect wealth. To this end, it interferes little in social or economic matters and leaves the market to its internal dynamics. To some extent, this has been related to the fact that the elite in politics also owned the capital necessary to trade economically and produce industrially. After World War I, the idea emerged to regulate industry and use the wealth thus generated. Essentially, the means/tools to implement regulation did not yet exist in the form they do today. Rather, there were no proven, tested and established tools to apply regulations on a large scale at that time. Therefore, the mechanisms used were based on what Marx and Engels had proposed some 50 years earlier: Governments should set priorities for the economic development of industry – which were the first steps towards the modern mechanisms of economic policy. The new state rulers no longer defined themselves as protectors of private property, but claimed state control over national economies. The approach of centralised control of the economy was viewed quite critically – liberal economic approaches were developed based on the argument put forward, for example, by Ludwig v. Mises in 1920, that saw logical impossibility in "socialist central planning" and called for a continuation of laissez-faire policies. These two directions basically determine the economic and consequently also the ecologic discussion to this day: the tension between non-interference and state influences on economic development and, more recently, the interaction between ecological issues and state control of the economy. Legislation can demand product characteristics that take ecological conditions into account and thus motivate innovations in this direction. It can also, for different reasons, accept the opposite and not regulate.

During the Great Depression, for example, a now controversial concept was developed that has had a strong influence on product design to this day: In order to shorten the life span of products, a deliberate end of life of products was proposed in order to create demand and thus stimulate the economy: the concept of "planned obsolescence". It was first introduced by Bernhard London in his book *Ending Depression by Planned Obsolescence* in 1932 [46], and still influences the design of products today. The concept has been implemented in several ways (Table 1.2).

Table 1.2: Implementation possibilities of the fail by design concept according to [46].

- Fail by design
- Design the product as not repairable/maintainable
- Convince customers that the product is significantly more valuable by offering new and slightly different features, thereby motivating them to discard the previously preferred product

If the concept is applied, planned obsolescence leads to reduced lifetimes, increased spending, not necessarily more wealth, but, assuming infinite resources, increased sales. The idea of intentionally limiting lifetimes to create more sales remained very much alive after the end of the Great Depression and is actually the opposite of the circular economy (CE) concepts discussed today.

To keep the economy growing, access to new markets became increasingly popular after World War II. What is understood as the development of the less developed countries was formulated in Harry S. Truman's inaugural address in 1949 [47]: "All countries, including our own, will derive great benefit from a constructive programme for the better use of the world's resources. Experience shows that our trade with other countries increases as they advance industrially and economically." Truman apparently implied that general progress for the multitude of people and cultures of the world should be found by emulating the material progress of the USA and its partners in what was then called the Free World. Wolfgang Sachs commented on the Truman Doctrine by stating that "Truman's imperative for development meant that Third World societies would be

- no longer seen as diverse and unparalleled opportunities for human life, but rather
- rather placed on a single 'track of progress', judged as more or less advanced according to the criteria of the Western industrialised nations" [48].

In essence, Truman's concept amounts to an internationalisation of the work of Adam Smith discussed above, for a market economy experiences rapid growth when productive capabilities expand. Essentially, according to this approach, Western cultures determine what standards to use and how to measure – the accepted unit is the ratio of GNP to population:

- Increased production is seen as the key to prosperity and peace – not well-being.
- Production, in turn, is seen as made possible by a wider and more vigorous application of modern scientific and technical knowledge.
- Although not explicitly formulated, free movement of capital is assumed – which also means consolidation of power structures to the advantage of the established.
- Diversity and a variety of development opportunities are not seen as a value. If considered at all, it is given away for international trade and a one-dimensional understanding of progress.

It was not until the mid-1970s that the realisation took hold that the state should act to protect the environment in addition to stimulating it in favour of rapid economic growth. The discussion was held both at universities and in the business community.

K. Boulding's 1966 essay *The Economics of the Coming Spaceship Earth* [51] is often cited as the origin of the concept of the "circular economy"(CE), although the essay does not contain this expression. The concept of CE was raised in 1989 by the British environmental economists David Pearce and R. Kerry Turner [52]: they pointed out that a traditional linear economy has been developed that has no built-in tendency to recycle. Essentially, under Adam's approaches, the economy treats the environment as an unlimited reservoir of resources and waste. In 1982, the self-replicating system was introduced by Walther Stahel. He speaks of "source cycles" and distinguishes between use, replenishment and production cycles [53]. Stahel's model of a value chain focuses on a spiral cycle system based on four activities:

1 reuse
2 repair
3 remanufacturing
4 recycling

The model is well comparable to today's understanding of circular economy.

A more quantitative approach was offered by Sankey [54]: Sankey diagrams are a special type of flow chart where the width of the arrows is shown proportional to the amount of flow. The figure on the next slides shows a Sankey diagram representing all primary energy flows in equivalents of CO_2 emissions. With this approach, a tool became available to quantify the pathways that materials travel during their use.

Figure 1.1 an example of a Sankey diagram for the use of paper in Europe. The figures refer to mass in kg [55].

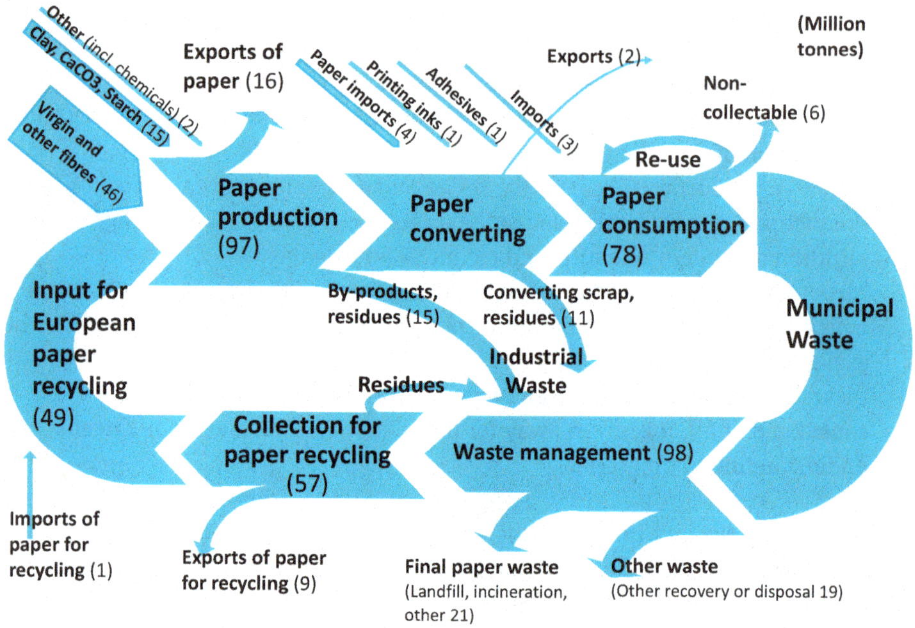

Figure 1.1: Sankey diagram for the use of paper in Europe. The figures refer to the masses in bio-kg. According to [55].

These approaches must be seen in the context of products that have also been developed with ecological criteria in mind. In 1962, Victor Papanek defined the responsible product design not as the design of the outer form. Papanek also talked about the responsible design in an ecological and social sense: "Design is fundamental to all human activities. . . . . Any attempt to separate design to make it a thing unto itself runs counter to the fact that design is the primary underlying matrix of life" [50]. The concept of responsible design was further developed by Ronald L. Mace, who stated in 1991: "Universal design is the design of all products and environments so that they can be used to the fullest extent possible by people of all ages and abilities" [51].

In the early 1990s, Tim Jackson began to assemble the scientific basis for CE – outlining this new approach to industrial production in his edited volume *Clean Production Strategies* [56].

In the early 1980s and in the aftermath of the 1968 movement, priorities such as those evident in the Truman Doctrine were also questioned by broader parts of the public. The model for wealth measured by GNP, which sketches the acceleration and deceleration of the economy based on the total of what was produced, implies a preference for large structures – but apparently does not take into account people's increasing/decreasing happiness ([58], see also Chapter 4). A different view was proposed by Ernst Schumacher in his book *Small Is Beautiful* [58], which was

highly acclaimed at the time and still provides valuable points of view today. According to Schumacher, the causes of the environmental crisis are due to:
1. modern technology – according to him, excessive mechanisation is inhumane and also kills creativity.
2. influence on the environment – he argues that producers and consumers do not care about the environment
3. depletion of non-renewable resources – according to him, economic growth can never be infinite because resources are finite.

Moreover, the way industry uses and implements modern technology neglects ethical aspects in particular, as it focuses on large-scale production, higher profits, non-renewable resources and economies of scale. According to Schumacher, this implies an inhumane and destructive economy that destroys the environment and the creativity of labour. To address the problems involved, Schumacher offers a number of characterising guidelines for solutions (Table 1.3).

Table 1.3: Labelling guidelines for solutions as outlined by Schumacher for his small is beautiful approach, after [59].

	Guidelines	Explanations
1	Small-scale production	Large-scale production is the cause of all environmental and social problems. Small scale is manageable and human. Therefore *"small is beautiful"*.
2	Ethical and normative	Use of ethical principles in production and consumption. *"The study of economics as if people mattered"*.
3	Human approach	Technology should be nonviolent and creative. *"Technology with a human face"*.
4	Environmental responsibility	All stakeholders should show responsibility towards environmental protection.
5	Intermediate technology	Based on renewable resources technology, labour intensive and creative. Schumacher says: Man is small, and so should produce on a small scale. "Small is beautiful".
6	Intermediate technology is/provides	Local employment, lower capital requirements, more creative, cheaper, based on local resources, not imported resources, no foreign dependence, local people are interested in protecting their own environment.

In the mid-1970s, and in response to the economic crisis of the time, policies in the major industrialised countries were driven by several concepts. For example, a strong liberalism in national programmes was geared towards supporting privatisation and limiting the influence of trade unions. At the same time, and still in the mindset of the Truman Doctrine, the free flow of capital to support production dominated international trade.

In 1971, Jay Forrester et al. published a paper on "World Dynamics" [60]. The mathematical models used allowed an examination of the interplay of several variables, in particular

- world population,
- food supply,
- world industrial production,
- pollution and
- resources (remaining).

In detail, the researchers' conclusion has essentially been that the idea of infinite growth in a finite world is inherently contradictory. These findings were taken further by the Club of Rome and Dennis Meadows, who used the two models in their publication entitled Limits to Growth in 1972 and 1976 (updated version). These studies strongly influenced public opinion and led to a change in opinion and are still to be considered as ground breaking today. The model calculations conclude that the world cannot support infinite growth. While the 1972 model led to a prediced collapse of the world around the year 2010, the model presented in a subsequent publication in 1976 pointed to a collapse before the year 2100. Moreover, the calculations showed that:

(i) Future world population, food production and industrial production will initially grow exponentially, become increasingly uncontrollable and then collapse during the twenty-first century.
(ii) The collapse will occur because the world economy will reach its physical limits in terms of non-renewable resources, agricultural land and the earth's capacity to absorb excessive pollution, all of which are finite.
(iii) Eleven vital minerals such as copper, gold, lead, mercury, natural gas, oil, silver, tin and zinc are depleted. If industrial production continues to increase, this too will lead to catastrophic results.
(iv) If current growth trends in world population, industrialisation, pollution levels, food problems and resource depletion continue unchanged, the limits to growth on this planet will be reached within the next hundred years. The most likely outcome predicted is a fairly sudden and uncontrollable decline in both population and industrial capacity sometime before the year 2010.
(v) Since technological progress cannot expand physical resources indefinitely, it would be prudent to set limits on our future growth rather than expect the end of the world within the next 50 or 100 years.
(vi) This catastrophe can be averted by controlling the rate of growth of production and population, reducing pollution and thus achieving a global equilibrium with zero growth.

The interactive simulation models of Limits to Growth produced a variety of scenarios that served primarily to define what should be prevented. It emphasised that

- pollution,
- high population growth rates and
- scarcity of food and resources

negatively affect the world's future prospects – which, as already predicted by Forrester, will lead to catastrophic outcomes for humanity:

Since resources are finite and likely to be depleted within 50 or 100 years, people should change their attitudes towards resource use, their own reproduction and pollution levels in order to save the world from collapse [61, 62].

So far, model prediction has provided surprisingly accurate predictions. A comparison between the predictions and the actual development can be seen in Figure 1.2.

The 1976 model can be considered a more optimistic model. With technological change, it predicts that a period of growth will initially occur in the future. However, growth will be followed by a decline in the birth rate, an increase in pollution and a decrease in the production of food, services and manufactured goods along with a decrease in available resources. The latter development has already begun: McKinsey describes the price development on the basis of its global commodity price index. According to McKinsey, and after a continuous decline between 1900 and about 2000, this value rises due to the increase in the price of raw materials. At the same time, price volatility for metals, food and non-food agricultural products was higher in the first decade of the twenty-first century than in any previous period [2, 20].

The Club of Rome's publication provided the discussion framework for the formation of a number of commissions, which, among other reasons, in turn prompted the UN to form a commission called "Our Common Future". This commission, also called the *Brundtland Commission*, after its head, wrote the "Report of the World Commission on Environment and Development: Our Common Future: From One Earth to One World" often called "The Brundtland Commission Report" [64]. The scope of the report in terms of environmental policy is intergenerational and global, seeking to reconcile the interests of poor and rich nations in terms of one. The report calls for a synthesis of the major issues: Nature, Society and Economy independent of economic growth. To address the issue, the Commission delivered the now classic call for development to be "sustainable to ensure that it meets the needs of the present without compromising the ability of future generations to meet their own needs".

It should be noted that Brundtland's definition of sustainability is anthropocentric. Rights of other living beings even if they impede human development are not conceded. This view of nature as a resource can, to some degree, be explained by the Commission's need to find a definition that was politically accepted by almost everyone at the time it was formulated – which risks missing the long-term goal of securing a place for humans in a natural habitat. On the other hand, this goal is

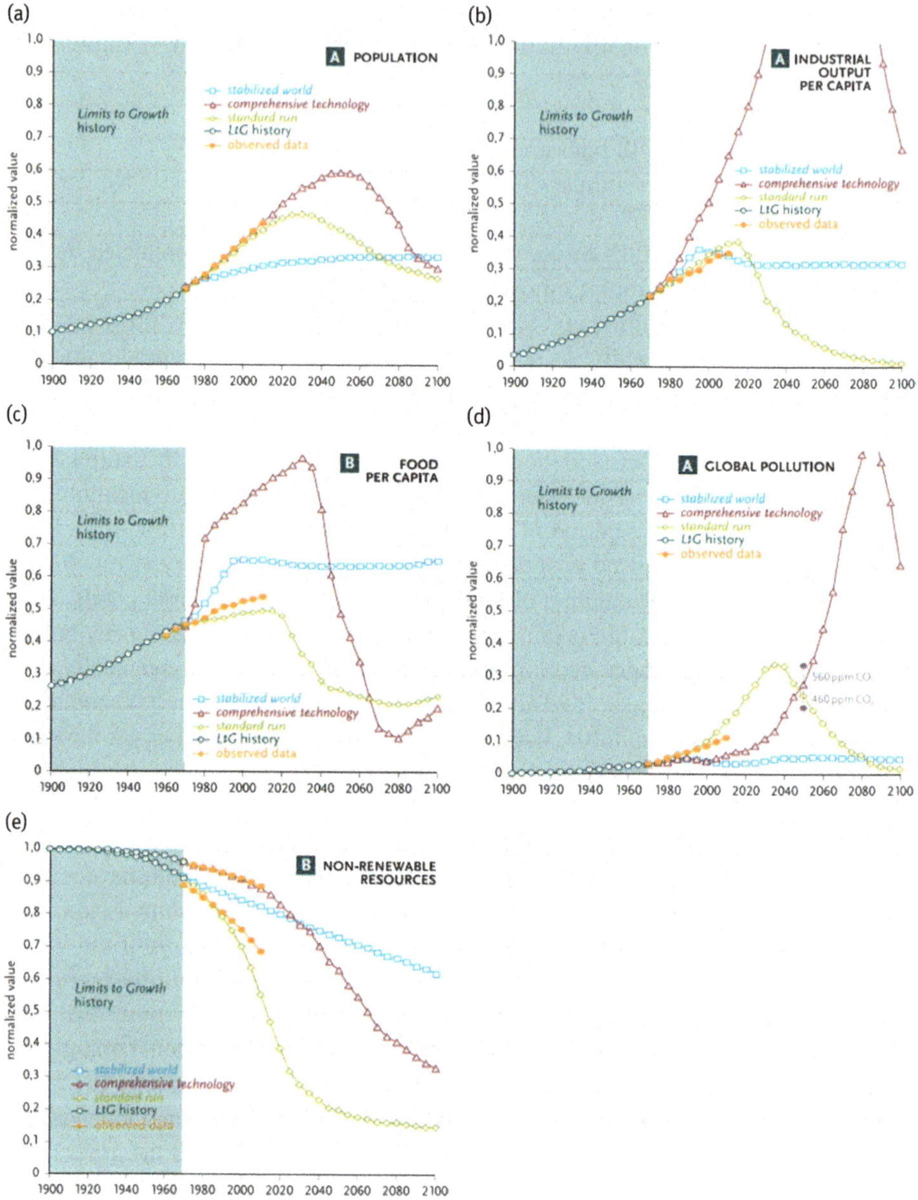

Figure 1.2: Club of Rome Limits to Growth (LtG) – predictions in 1976 and comparison with actual values in 2014. Comparison of historical data with three Limits to Growth scenarios, for population, industrial output per capita, food per capita, global persistent pollution, fraction of non-renewable resources remaining. Based simply on the comparison of observed data and the Limits to Growth scenarios, the standard run scenario seems to better align with the observations made than the other two scenarios. It is remarkable that the model assumptions for the straightforward model (standard run) seem to allow good predictions [63].

based on the fundamental assumption that nature itself is stable and sustainable by providing a safe place for species to survive. But this is not the case – nature is not stable from the point of view of individual ecological niches, strictly speaking not sustainable in the original sense of the term:

It is obvious that even before the existence of humans, species arose and disappeared again. There was no guarantee that survival of future generations was possible for a particular species – in the same habitat or elsewhere. Rather, it was (and is) the case that nature is characterised by constantly changing local equilibria between ecological niches that offer no guarantee of stability, let alone the survival of a particular species.

Nevertheless, the Brundtland Commission's definition sets new standards for what is called development, and it changes the view of the definition as given in the Truman Doctrine. Measured against these standards, not all Western cultures can be considered truly developed.

The work of the commissions set the framework for a real shift in the accepted ground truths (paradigms) (for a definition of this, see [65]). The Brundtland Report changed the assumptions that dominated the previous approach to economic development. The concept of sustainable development is presented as a serious response to the challenges facing the world in terms of limits to growth and the global crisis as interlocking natural, social and economic challenges. The Brundtland Report emphasises that ecology and economy are interlinked and embedded in society and must be thought of together. The report states ([64], p. 5): "Ecology and economy are increasingly intertwined – locally, regionally, nationally and globally – into a seamless web of causes and effects." It goes on to highlight that population is linked to other issues such as women's empowerment ([64], p. 11) or tribal and indigenous peoples' right to livelihoods, an alternative term for economy. It is made clear that what is commonly understood as development is made responsible for the

- Destruction of the environment
- Growing inequality
- Growing debt burden of poor nations
- Lack of attention to the common commons (common goods)
- Lack of economic diversification at local and regional levels.

Following the publication of the Brundtland Report in 1987, the first Earth Summit was organised in Rio in 1992. The outcomes triggered by the report and addressed at the Rio Summit were legally binding agreements (Rio Convention [66]).

The Rio Convention in turn led to a number of legislative initiatives. For example, the European Commission defined specific requirements for the eco-design of products [67] (specific requirements in the sense of measured parameters and limit values – for example, a limit value for the amount of energy that products may consume or a minimum content of recycled material that must be used in production).

Table 1.4: Binding agreements of the Rio Convention.

– Convention on biological diversity (measures against the destruction of natural ecoregions and so-called uneconomic growth) – Framework Convention on Climate Change (UNFCCC) (which in turn led to the Kyoto Protocol and the Paris Agreement) – United Nations Convention to Combat Desertification – The agreement "not to conduct activities on indigenous peoples' lands that cause environmental degradation or are culturally inappropriate"

There are also generic requirements that do not set a limit value but require, for example, that

- the product is "energy efficient" or "recyclable",
- the seller provides information on how to use and maintain the product to reduce its environmental impact,
- maintenance may be enhanced to minimise the products environmental impact by extension of lifetime,
- the seller carries out a life cycle assessment of the product in order to identify alternative to identify alternative design options and potential improvements.

The introduction of new minimum requirements can lead to non-compliant products no longer being allowed to be sold in EU countries. An example of this is traditional light bulbs, which were phased out from 2009. Finally, the EU has also set a path for the transition to CE as a goal [67].

It aims to transition to an economy in which the value of products, materials and resources is retained in the economic cycle for as long as possible and their loss from the cycle as waste is minimised. In this context, a global measure of the use of natural resources is helpful – for this purpose, the Earth's carry-on-capacity was introduced (Figure 1.3), which describes how many Earths are needed to sustainably satisfy the consumption of raw materials in the long term. This indicator shows very clearly how far the overexploitation of existing resources has progressed.

One aim of the legislation was to contribute to energy saving and to avoid irreversible damage caused by the consumption of resources. Moreover, in the long term, the measures described above are expected to foster a sustainable, low-CO_2-emitting and resource-efficient economy, which, it is hoped, will in turn generate sustainable competitive advantages for Europe.

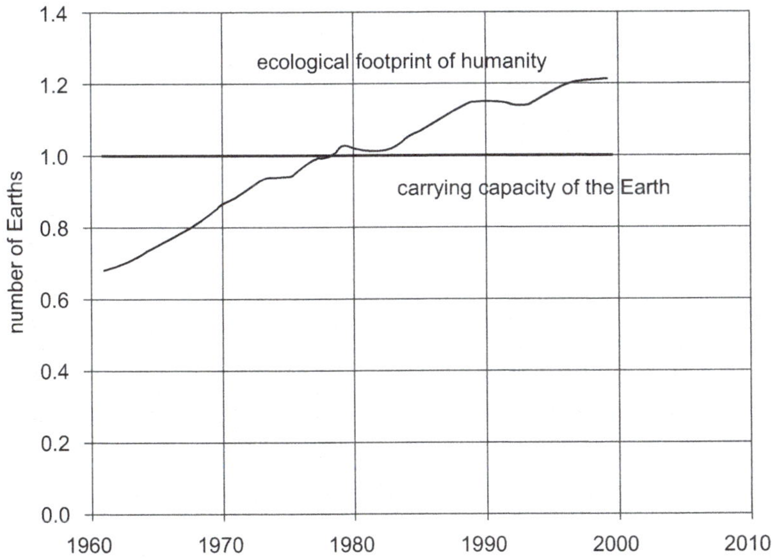

Figure 1.3: Carry-on-capacity of the Earth according to [68].

References

[1] https://edition.cnn.com/2019/07/03/asia/india-heat-wave-survival-hnk-intl/index.html

[2] www.theguardian.com/commentisfree/2014/sep/02/limits-to-growth-was-right-new-research-shows-were-nearing-collapse

[3] Global warming of 1.5 °C, An IPCC Special Report on the impacts of global warming of 1.5 °C, October 2018 by the IPCC, Switzerland. IPCC website www.ipcc.ch, ISBN 978-92-9169-151-7

[4] Unclear climate targets, Prof. Reto Knutti, 08.12.2015 (->11)

[5] https://unfccc.int/process-and-meetings/the-paris-agreement/the-paris-agreement) and follow the related path of action (https://www.ipcc.ch/sr15/

[6] Global warming of 1.5 °C, An IPCC Special Report on the impacts of global warming of 1.5 °C, October 2018 by the IPCC, Switzerland. Electronic copies of this Summary for Policymakers is available from the IPCC website www.ipcc.ch, ISBN 978 92 9169 151-7

[7] Existential climate-related security risk: A scenario approach, Melbourne, May 2019, David Spratt & Ian Dunlop, Chris Barrie, National Centre for Climate Restoration, https://breakthroughonline.org.au

[8] https://www.ethz.ch/en/news-and-events/eth-news/news/2015/12/unclear-climate-targets%20.html,https://www.ethz.ch/content/main/en/news-und-veranstaltungen/eth-news/news?AUTHOR=UHJvZi4gUmV0byBLbnV0dGk&path=L2NvbnRlbnQvbWFpbi9lbi9uZXdzLXVuZC12ZXJhbnN0YWx0dW5nZW4vZXRoLW5ld3MvbmV3cy9qY3I6Y29udGVudA, Unclear climate targets, 08.12.2015

[9] https://advances.sciencemag.org/content/advances/3/8/e1603322.full.pdf

[10] https://www.bmz.de/rue/en/concepts_topics/GFG_Extractive_Sector/eiti/index.html

[11] IARIGAI Conference Stuttgart, Sept 2019, Designing an industrial product following criteria for circular economy. What product designers should consider – a case study based on a printing press, Michael Has, International school of paper, print media and biomaterials, 461 Rue de la Papeterie, 38402 Saint-Martin-d'Hères, France
[12] https://www.nature.com/articles/s41586-018-0071-9
[13] www.ilo.org/weso-greening/
[14] https://www.nytimes.com/2018/06/29/opinion/sunday/immigration-climate-change-trump.html
[15] https://www.rohma.ch/de/regulierung/rohstoffgesetz-rohg/, Schweizerisches Rohstoffgesetz ROHG, 1. April 2014;Auswirkungen der Klimaänderungen auf die Schweizer Volkswirtschaft, Schweizerisches Bundesamt für Umwelt, 2007, P 83
[16] https://www.rohma.ch/de/regulierung/rohstoffgesetz-rohg/, Schweizerisches Rohstoffgesetz ROHG, 1. April 2014;Auswirkungen der Klimaänderungen auf die Schweizer Volkswirtschaft, Schweizerisches Bundesamt für Umwelt, 2007, e.g. p 104
[17] https://www.thebalance.com/why-are-food-prices-rising-causes-of-food-price-inflation-3306099
[18] https://climateandsecurity.org/2017/11/22/defense-bill-passes-with-climate-change-and-national-security-provision/
[19] https://www.wsj.com/articles/how-companies-are-pushing-ahead-on-climate-change-targets-1510790610?shareToken=steffd42162ba04cc2985157b82bcdaea3
[20] Grilli and Yang; Pfaffenzeller; World Bank; International Monetary Fund; Organisation for Economic Cooperation and Development (OECD) statistics; Food and Agriculture Organization of the United Nations (FAO); UN Comtrade; McKinsey Global Institute analysis
[21] EU Directive WEEE (2012/19/EU)
[22] https://ec.europa.eu/clima/policies/strategies/2030_en
[23] German "Verpackungsgesetz" as per 1. January 2019, §21 (1) 2 and § 16 (Bundesanzeiger Verlag (2017): Gesetz zur Fortentwicklung der haushalts-nahen Getrennterfassung von wertstoffhaltigen Abfällen. http://www.bgbl.de/xaver/bgbl/start.xav?startbk=Bundesanzeiger_BGBl&jumpTo=bgbl117s2234.pdf
[24] Global trends in climate change litigation: 2019 snapshot, Joana Setzer and Rebecca Byrnes, Policy report; July 2019 www.lse.ac.uk/GranthamInstitute/climate-change-laws-of-the-world/
[25] Climate change litigation: A new class of action; https://www.whitecase.com/publications/insight/climate-change-litigation-new-class-action"
[26] The Status of Climate Change Litigation: A Global Review, United Nations Environment Programme; Columbia University, Sabin Center for Climate Change Law (2017-05) https://wedocs.unep.org/bitstream/handle/20.500.11822/20767/climate-change-litigation.pdf?sequence=1&isAllowed=y
[27] Swiss Commodity Market Supervisory Authority Act, CMSAA, of 1 April 2014 or the Commodities Act, CA, of 1 April 2014; https://www.rohma.ch/en/regulation/commodities-act/
[28] Should trees have standing? Towards legal rights for natural objects (C. Stone). https://iseethics.files.wordpress.com/2013/02/stone-christopher-d-should-trees-have-standing.pdf
[29] https://www.smithsonianmag.com/smart-news/toledo-ohio-just-granted-lake-erie-same-legal-rights-people-180971603/#Vc7oA942qs1VfaLI.99
[30] Framework Directive 2005/32/EC
[31] www.ilo.org/weso-greening/
[32] www.nytimes.com/interactive/2017/02/17/world/americas/mexico-city-sinking.html
[33] www.thebalance.com/department-of-defense-what-it-does-and-its-impact-3305982
[34] https://www.lawphil.net/judjuris/juri1993/jul1993/gr_101083_1993.html

[35] https://www.economist.com/open-future/2019/09/17/make-a-healthy-climate-a-legal-right-that-extends-to-future-generations?utm_source=Nature+Briefing&utm_campaign=9cccfcc954-briefing-dy-20190919&utm_medium=email&utm_term=0_c9dfd39373-9cccfcc954-43827413
[36] www.labelandnarrowweb.com/contents/view_breaking-news/2019-09-18/flint-cps-inks-expects-future-price-increase/- accessed 10.10.2019; 16:31 h
[37] www.britannica.com/event/Industrial-Revolution
[38] Smith, A., An Inquiry into the Nature and Causes of the Wealth of Nations, London, 1776; See Smith, A. (1776). https://books.google.com/books?id=C5dNAAAAcAAJ&pg=PP7 "\l"v=onepage&q&f=true. London: W. Strahan. 1 ed., Retrieved April 11, 2020
[39] Malthus, T., An Essay on the Principle of Population, https://archive.org/details/essayonprinciplOOmalt (1 ed.). London: J. Johnson in St Paul's Church-yard. 1798. Retrieved April 11, 2020
[40] Ricardo, D., On the Principles of Political Economy and Taxation, Murray, J., Albemarle-Street, London, 1817
[41] Marx, K., Engels, F. Manifesto of the Communist Party, London, 1848 (Marx/Engels Selected Works, Vol. One, Progress Publishers, Moscow, 1969, pp. 98–137;)
[42] Charles, E.Z., in https://www.sciencedirect.com/science/referenceworks/9780080970875, 2015
[43] Meindertsma, C., Vitra Design Museum, November 2018; https://www.design-museum.de/de/ausstellungen/detailseiten/christien-meindertsma-beyond-the-surface.html Retrieved April 11, 2020
[44] Bartels, R., The History of Marketing Thought, Columbus, Ohio, Grid, 19761965, 1976
[45] Demirdjian, Z.S., "Rise and Fall of Marketing in Mesopotamia: A Conundrum in the Cradle of Civilization," In The Future of Marketing's Past: Proceedings of the 12th Annual Conference on Historical Analysis and Research in Marketing, Leighton Neilson (ed.), CA, Longman, Association for Analysis and Research in Marketing, 2005; Clarke, J.R., Dobbins, J.J., Foss, P.W., The World of Pompeii, Oxford, Routledge, 2008, p. 330; Curtis, R.I., "A Personalized Floor Mosaic from Pompeii," American Journal of Archaeology, Vol. 88, No. 4, 1984, DOI: 10.2307/504744, pp. 557–566, STabelle URL: https://www.jstor.org/sTabelle/504744
[46] Bernhard London, Ending Depression by Planned Obsolescence, New York, 1932
[47] Truman, H.S., Inauguration address; 1949; https://www.bartleby.com/124/pres53.html: Retrieved April 11, 2020
[48] Sachs, W., The archaeology of the development idea, www.burmalibrary.org/docs14/The_Archaeology_of_the_Development_Idea.pdf; Retrieved April 11, 2020
[49] Papanek, V. 1962 Design für die reale Welt. Anleitungen für eine humane Ökologie und sozialen Wandel, 1962; https://archiv.kultur-punkt.ch, Retrieved April 11, 2020
[50] Ronald, L. Mace quoted in http://www.design.ncsu.edu/cud/about_ud/udhistory.htm Retrieved April 11, 2020
[51] Kenneth, E. Boulding: http://arachnid.biosci.utexas.edu/courses/THOC/Readings/Boulding_SpaceshipEarth.pdf. In: Jarrett, H. (Hrsg.): Environmental Quality in a Growing Economy, Essays from the Sixth RFF Forum on Environmental Quality. Baltimore: The Johns Hopkins Press 1966. S. 3–14.
[52] Pearce, D., Turner, R.K. Economics of Natural Resources and the Environment, Johns Hopkins University Press, 1989. https://en.wikipedia.org/wiki/International_Standard_Book_Number "\o"InternationalStandardBookNumber; 978-0801839870.
[53] Stahel, W., The Product-Life Factor, 1982; https://Product-life.org/en/major-pub-lications/the-product-life-factor, Retrieved April 11, 2020
[54] Sankey, H.R. The Thermal Efficiency of Steam-Engines. In: M.p.i.c.e. 125, 1896, S. 182–242.

[55] Has, M. et al, Design and Management for Circularity – the Case of Paper; World Economic Forum Davos 2016; http://www.cepi.org/publication/design-and-management-circularity-%E2%80%93-case-paper, Retrieved April 11, 2020
[56] Jackson, T., Circular Economy edited collection Clean Production Strategies, 1990; https://www.theseus.fi/bitstream/handle/10024/113985/Sudentaival_Hanne_Introduction%20to%20Circular%20Economy.pdf?sequence=1, Retrieved April 11, 2020
[57] Bhutan's Gross National Happiness Index, Oxford poverty and Human Development Initiative; https://ophi.org.uk/policy/national-policy/gross-national-happiness-index/ Retrieved April 11, 2020
[58] Schumacher, E.F.; Small Is Beautiful: Economics as if People Mattered; www.daastol.com/books/Schumacher%20(1973)%20Small%20is%20Beautiful.pdf, Retrieved April 11, 2020
[59] Schumacher, E.F.; Small Is Beautiful: Economics As If People Mattered: 25 Years Later . . . With Commentaries (1999). ISBN; https://en.wikipedia.org/wiki/Hartley _%26_Marks_Publishers" "Hartley & Marks Publishers
[60] Jay Forrester, World Dynamics. Wright Allen Press, Cambridge, USA, 1971, https://monoskop.org/File:Forrester_Jay_W_World_Dynamics_2nd_ed_1973.pdf, Retrieved April 11, 2020
[61] http://www.peakoilindia.org/wp-content/uploads/2013/10/Limits-to-Growth-updated.pdf, Retrieved April 11, 2020Retrieved April 11, 2020
[62] https://insightmaker.com/insight/1954/The-World3-Model-A-Detailed-World-Forecaster, Retrieved April 11, 2020
[63] Turner, G.M., On the Cusp of Global Collapse? GAIA 21/2(2012): 116–124
[64] Gro Harlem Brundtland "Report of the World Commission on Environment and Development: Our Common Future: From One Earth to One World" http://www.un-documents.net/wced-ocf.htm, Retrieved April 11, 2020
[65] Thomas, S. Kuhn: The Structure of Scientific Revolutions, University of Chicago Press, 1970
[66] Rio Earth Summit in 1992 (Rio Convention) www.un.org/en/development/desa/population/migration/generalassembly/docs/globalcompact/A_CONF.151_26_Vol.I_Declaration.pdf, Retrieved April 11, 2020
[67] Eco-Design requirements of the EU, europa.eu/youreurope/business/product/eco-design/index_en.htm Retrieved April 11, 2020
[68] B. Murrey as quoted in http://www.wrap.org.uk/sites/files/wrap/Embedding%20sustainability%20in%20design%20%20-%20final%20v1.pdf, Retrieved April 11, 2020

Chapter 2 Economic-ecological concepts

2.1 Value creation and supply chains

A company exists to create value – and to find new ways to create even more value. Since a company always exists in the context of supplying companies (which also create value) and supplied customers, one speaks of *value chains* in this context. Basically, a value chain consists of a series of activities that are carried out to create value – be it in one or more companies. The elements in value chains are interconnected – hence the term "chain" – and often entire chains are interlinked, which in practice makes them difficult to delineate. According to Michael Porter, a value chain focuses on systems and the primary and supporting activities (Table 2.1) within them [1].

Table 2.1: Primary activities in value chains according to [1].

1. Inbound logistics (goods receipt, warehousing and inventory management of raw materials)
2. Operation (efforts to transform raw materials into a product)
3. Delivery logistics (includes logistics after operations are completed and the product is made available to the customer)
4. Marketing and distribution (including related strategies, channel selection, promotion and pricing)
5. Service and maintenance (activities to create and improve customer experience, including refund and exchange programmes and warranty)

All five of the above-mentioned activities aim to create an advantage that can be sold to a potential customer. The documented activities determine the costs, influence the profit and thus the success, which in turn is measured as margin.

In this concept, a value chain is seen as linear. In analogy to the concept of circular economy – see below – a linear value chain is replaced by a value cycle. However, the concept can be extended in a circular way. The intention of the concept is that products are returned into the value chain after production and reused (as a product, part or material) or start a new life in a new context – but do not become waste. To make such a concept effective, one has to take into account the internal roles of the company as well as all actors in a supply chain and interacting supply chain and their respective value contributions – which is by no means self-evident when more than two companies in combination are involved in a value chain and want to benefit from it, because these companies do not benefit under all conditions from the value of the innovation they bring into the value chain.

https://doi.org/10.1515/9783110767308-002

2.1.1 Supply chain

The term "supply chain" is often used in the context of value chains and has its origins in solving the problem of supporting troops in combat. A supply chain is created through specialisation and through a non-local supply of raw materials. It refers to the sequence of actions and logistics in production. Most often, a supply chain involves more than one company. The structure of supply chains can therefore be complex and often not static, because ideally the participants are interchangeable and are changed accordingly to reduce costs.

It is important to note that there is not just one supply/value chain for one type of product. Rather, a supply chain can take many different forms, often enabling new value creation and opportunities. In closed-loop supply chains, in addition to the typical forward flows, there are also reverse flows of used products (post-consumer use) back to the manufacturers. Examples are supply chains with post-consumer-use returns or leasing, and end-of-use returns with re-manufacturing [2].

As mentioned, a supply chain based on the specialisation of individuals involves the delivery of raw materials, their preparation, storage and transport. This effort is accompanied by an environmental footprint [3].

Supply chains also come with risks, as there is a link between increased material footprint and security of supply – for example, if a company becomes dependent on a supplier who or whose product has a high carbon footprint. McKinsey points out in this context that relatively few companies work with their suppliers to manage risk [4]. To help companies identify the critical sustainability issues in supply chains, several organisations have developed measurement frameworks and tools:

- The Sustainability Consortium developed a set of performance indicators and a reporting system to identify and highlight sustainability hotspots for many consumer goods categories – covering 80–90% of consumer goods impacts, according to McKinsey [5].
- The World Wildlife Fund offers more than 50 performance indicators to measure supply chain risks and the likelihood and severity of those risks associated with the production of various goods [6].
- The Sustainability Accounting Standards Board has developed standards to help listed companies in 10 industries, including consumer goods. The aim is to provide investors with essential information on the sustainability performance of companies along the value chain [7].
- CDP and the Global Reporting Initiative have created standards and metrics for comparing different types of footprints.

2.1.2 Technological progress in value chains – a dilemma

There are various impulses for innovation – e.g. practical working conditions and new (technical) solutions, but also legal changes that in turn require new solutions (e.g. reduced legal limits for emissions). In circular value chains, there is a new dynamic when it comes to innovation: different actors are involved in a value chain, be it linear or circular. These actors form an economic network, possibly without being in a direct contractual relationship.

From an innovation perspective, an interesting situation arises in which one party benefits from innovations of another party without a direct business relationship between two parties. In such a situation, the added value created by the innovation does not provide added value to the direct business partner. If the intervening party or parties have no added value or would even have to invest due to the change in conditions, they are unlikely to do so, as there is neither a justification for such investments nor an incentive for one party to implement the innovation. While this problem can be overcome when there is only one party between the companies involved, it becomes more difficult to implement the innovation the further apart the innovator and the beneficiary are. Due to the increased attention to sustainability and the accompanying changes in part of the circular value chain, this situation is occurring more frequently and therefore needs to be addressed accordingly.

Three possible approaches to overcome this situation are summarised in Table 2.2.

Table 2.2: Ways out of the dilemma of innovation in value chains.

- Brand owners can get a detailed overview of all steps in the supply chain and the associated prices and processes. If they use their procurement power, they can and will push through changes, even if there may be higher costs for the individual party making the changes. This can also favour changes that lead to a reduced environmental footprint of the whole cycle.
- Legal requirements can force changes in the supply chain – leading to changed costs in terms of monetary value.
- New players can enter the market and change rules of the game, especially by integrating other players – especially when new legal conditions come into play.

2.2 Circular economy

In linear value chains, products are designed to fulfil their purpose before they become waste after use. From the customers' or users' point of view, a long life span is usually desirable, while from the manufacturers' point of view, a short life span is of interest, as it is likely to contribute to higher sales. However, both perspectives usually imply a single lifetime of a product – a concept often paraphrased as "cradle-to-grave" [12]. Given the finite nature of natural resources, such a concept cannot be recommended easily.

In contrast, the "cradle-to-cradle" concept [12] aims to overcome the limitation of the cradle-to-grave approach by planning for both minimised displacement and prolonged use of raw materials when designing a product, which can be achieved, for example, by reusing many components. Implicit is the need for a more or less sophisticated mechanism for disassembling the product into its component parts at the end of its life, separating and testing these parts and returning them to subsequent production cycles. This goes hand in hand with returning the recyclable components (parts, raw materials) to the supplier and developing profitable supply chains for this. In addition, the consumption of energy and materials during production and the generation and emission of hazardous waste throughout the cycle must be minimised.

Against this background, a circular economy is an economic system in which products are produced, traded, used and recycled in closed cycles. A circular economy is characterised as an economy that is regenerative by design: this with the aim of retaining as much value as possible from products, both parts and materials [9]. In a circular economy, material cycles are closed along the lines of natural ecosystems; toxic substances are eliminated and ideally no waste is produced, as all residual flows are valuable as a resource and are reused. In addition, products are taken back after use for repair and remanufacturing in order to use parts or the whole products a second, third or several times, and residual streams are separated in a biological and technical cycle [10]. Circular economy is a regenerative economic system. The business models involved are not based on the traditional 'end-of-life' concept but rather focus on reduction of consumption, recycling and recovery of parts and materials throughout all phases of a product's lifetime. The reason for these priorities is to spend less in terms of environment and money, thereby achieving a more sustainable development that permits increased environmental quality in combination with economic prosperity.

The methods for doing this are reducing the need for materials and the resulting waste, reusing products (and parts) and recycling materials. The circular economy refers to a deliberately sustainable industrial model in which products are designed to

- reuse,
- disassembly,
- remanufacture and
- recycle.

The intention is that a large proportion of materials are reused rather than produced through primary extraction – also to save energy. The model requires that resources be kept in use for as long as possible so that maximum benefit can be derived from them, followed by the recovery and regeneration of raw materials once the manufactured goods have reached the end of their life [11].

However, it must be remembered that all consumption requires the use of energy for the distribution of both the product and the raw materials, and reuse of any kind similarly contributes to energy requirements. It follows that recycling also requires energy and will therefore never be 100% regenerative.

Against this background, circular economy intends to replace the end-of-life concept with

- recovery,
- conversion to the use of renewable energy,
- the elimination and renunciation of the use of toxic chemicals, the reuse and recycling back into the biosphere and
- minimising or eliminating waste through better design of materials, products, systems and business models.

To replace an end-of-life approach, the production and use system must necessarily be restorative and regenerative at all stages by intention and design [11].

References

[1] Porter, M.E. Competitive Strategy. Techniques for Analyzing Industries and Competitors. Free Press, New York 1980; https://www.albany.edu/~gs149266/Porter%20(1985)%20–%20chapter%201.pdf, Retrieved April 11, 2020.

[2] Souza, G.C. a critical review of closed loop Supply chain: A critical Review, and Future Research. Decision Sciences, 44(1), 2012.

[3] https://www.mckinsey.de/api/sitecore/InteractiveWrapper/RenderView?sc_itemid=%7B0237E967-A10A-489F-B428-C5AA3437D98F%7D, Retrieved August 24, 2021.

[4] www.mckinsey.com/business-functions/sustainability/our-insights/starting-at-the-source-sustainability-in-supply-chains Retrieved April 11, 2020.

[5] https://www.sustainabilityconsortium.org/2019/09/tsc-impact-report-shows-30-improvement-in-consumer-goods-supply-chain-transparency-since-2016/, Retrieved April 11, 2020.

[6] https://www.worldwildlife.org/blogs/sustainability-works?blog_category_id=supply-chain-management, Retrieved April 11, 2020.

[7] https://www.sasb.org/, Retrieved April 11, 2020.

[8] https://www.cdp.net/en/supply-chain Retrieved April 11, 2020.

[9] Ellen Mac6 Oxford Business School, 2016, https://www.sbs.ox.ac.uk/oxford-answers/what-circular-economy, Retrieved April 11, 2020.

[10] Ellen MacArthur Foundation, 2015a e.g. Promoting Remanufacturing, Refurbishment, Repair, and Direct Reuse As a contribution to the G7 Alliance on Resource Efficiency, Brussels, Belgium, 2017, https://ec.europa.eu/environment/international_issues/pdf/7_8_february_2017/workshop_report_Brussels_7_8_02_2017.pdf, Retrieved April 11, 2020.

[11] Ellen MacArthur Foundation, Economic and Business Rationale for an Accelerated Transition; Cowes, Isle of Wight, 2012).

[12] McDonough, W., Braungart, M. (2002). Cradle to Cradle: Remaking the Way We Make Things. New York: North Point Press. ISBN 0865475873. OCLC 47623923.

Chapter 3
Business and sustainability

3.1 Business models

Business models explain what a company will sell (products or services), how it intends to market that product or service, the price (and cost) of a product and how the activities will be turned into profit. However, the term business model is not uniformly defined [1], different definitions agree that a business model is a holistic approach that examines the activities of a company (including upstream and downstream processes) and the creation of value [2], which also combine ecology with social, financial and cultural aspects of doing business [3].

Due to the diversity of business types and changing market situations, there is not one business model, but a great variety. There are generic types of business models such as indirect or direct sales, retail, rental, leasing, software as a service (hardware or software), consulting, interim management, food farming, franchising, advertising based and brick-and-mortar businesses, to name just a few [4, 5]. There are also hybrid models that mix aspects of two or more of these models, for example, by combining internet commerce with brick-and-mortar shops.

Business models use collected information to explain what a company will sell (product or service), how it intends to market that product or service, the price (and cost) of a product and how the activities will be turned into profit [6–9].

The earliest business models were very straightforward and little more than a simple plan for how the business was going to make money. Today's business models are often more sophisticated and typically involve the interaction between the customer and the producer (including the supply and value chain), the attraction of investment, the ability to attract and retain motivated talent and employees and, for example, how assets are managed. Understanding business models today is based on a complex mosaic of individual and interacting models.

3.2 Sustainable business models

There are different definitions of what a business model for sustainability is (for a discussion, see [10]). According to this source, "a sustainability business model helps to describe, analyse, manage and communicate

(i) a company's sustainable value proposition to its customers and all other stakeholders,
(ii) how it creates and delivers this value,
(iii) and how it captures economic value while maintaining or regenerating it, social and economic capital beyond its organisational boundaries."

https://doi.org/10.1515/9783110767308-003

The sustainable value proposition of a product or service is reflected in the material value achievable after use and implicitly in its price. The desire to buy is, most often, not driven merely by vital needs but also by values reflected in or represented by the product – seeking out a particular value proposition for specific target customers or groups. Sirgy and Danes [11] describe the desire to buy through the desire to identify with the product or the values it represents. This connection is used by companies actively or supposedly committed to achieving social and environmental goals, because this attracts target groups that want to be perceived as acting sustainably to the corresponding products, which in turn can support a higher price point in the market.

According to Lüdeke-Freund, the main purpose of a business model is "defined as the preservation or regeneration of natural, social and economic capital by offering a sustainable value proposition to customers and other stakeholders" and "places particular emphasis on the main functions of the business model, namely describing, analysing, managing and communicating" [12]. The development and implementation of business models for sustainability and sustainable products are closely linked [13]. By aligning a business model with the principles of the circular economy, it is possible to add value to product development – for example, by increasing the residual value of products after their useful life to support recyclability and residual value at the end of the life cycle as the product, assemblies or materials can be better reused.

Many models and theories on sustainable business models have emerged in research in recent years. The theories discussed in the literature in this context include market supply, customers and resources, so that sustainability-related issues can be included in the existing network of models provided [22]. Wirtz's approach [14] is suitable for the inclusion of sustainability-related issues. Resources and market supply (in Wirtz's terminology generally: the inbound and outbound situation) are included in the form of carbon footprints and material footprints (see below). Including the CO_2 footprint in a business model is not obvious, because while the material price reflects the equilibrium of a cost- and demand-driven market for materials, the price of CO_2 emissions has to be calculated on the basis of the consumption of energy and of the material used.

(a) The calculation scheme is standardised – which is achieved via the GHG emission calculation approaches (see respective chapters below).
(b) The price related to emissions to the costs of the already visible and known effects of CO_2 emissions or alternatively based on the cost of the expected impact of CO_2 emissions on a (global) scale.

Both approaches would lead to different prices – many countries and economic regions have adopted a way of pricing CO_2 emissions that is reflected in the price of the product (see respective chapters below). In 2020, the price range is from a few (in the USA) to about €140 per tonne of emitted CO_2 in Sweden. The additional price contribution resulting from sustainability aspects could be levied as an environmental tax or through the value chains starting with the original suppliers of raw materials (key industries) and energy.

3.3 Schools offering theories on ecologically sustainable business models

In the 1970s, the need for an integrated view of nature, people and the economy became apparent to a broader public. It became clear that a holistic approach was needed to establish an economy that minimised the consumption of natural resources. The complexity of interactions in a system that includes crops, trade, biosphere and pollution would also have to be taken into account to enable the economy to function efficiently without neglecting the needs of nature. Since then, the idea of sustainability has been further developed in business models, with several schools of thought emerging. Fundamental to all is the idea that the business model should promote a reduction in waste, emissions and material consumption per use.

Different schools have developed different variants of the circular economy, which have in common that they propose an economic model that is clearly different from linear economic activity (Table 3.1).

Table 3.1: Different schools interpreting the circular economy in different ways (see detailed descriptions below).

Originator(s), name of model, source	Flavour of circular economy
K. E. Boulding, 1966, **Earth as a spaceship**	Proposed in idea of circular flows of materials as a model for economy
Michael Braungart, William McDonough **Cradle-to-cradle**	Closed-loop concept, illustrated by the expression *cradle-to-cradle* (in opposition to cradle-to-grave)
Janine Benyus, 2002 **Biomimicry**	Studying natural processes as models to be replicated in human activities; inspired the idea of imitation of natural cycles capable of continuous regeneration
G. Pauli **Blue economy proposal**	Follows the analogy of ecosystems, which create neither waste nor emissions and supply themselves with local resources only
Walter Stahel **Performance economy**	Focus is on the maintenance and exploitation of stock (mainly manufactured capital) rather than linear or circular flows of materials or energy – long term the concept leads economy to become service-based
Amory Lovins **Natural capitalism**	Considers the influencing forces, determines economy and adds the supply of natural capital as a key ingredient
Thomas Grädel **Industrial ecology**	Aims to develop industrial processes that minimise material waste and pollutants in materials in line with the cradle-to-cradle concept
P. Laybourn **Industrial symbiosis or industrial metabolism**	Sustainable management (by quantity and quality) of materials and energy entering and leaving production processes

The three core principles common to the different approaches to the circular economy are given in Table 3.2.

Table 3.2: Core principles of the circular economy.

– Preserving and enhancing natural capital by controlling stocks and balancing renewable resource flows; optimising resource use and preserving value along the entire value chain
– Optimising resource yield by circulating products, components and materials with the highest utility at all times in technical and biological cycles; downstream value chains are linked to upstream value chain
– Promoting system effectiveness by identifying and designing out negative externalities; systemic impacts are identified, understood and potentially mitigated and considered along with the total value of the product produced, sold and taken back for reuse

If those principles are followed throughout the phases of the life of a product, the value of materials is ideally not destroyed. In cases where current technology does not allow for the full recovery of all such materials, the non-recoverable components should at least not destroy the value of the materials for which recovery is possible (Figure 3.1).

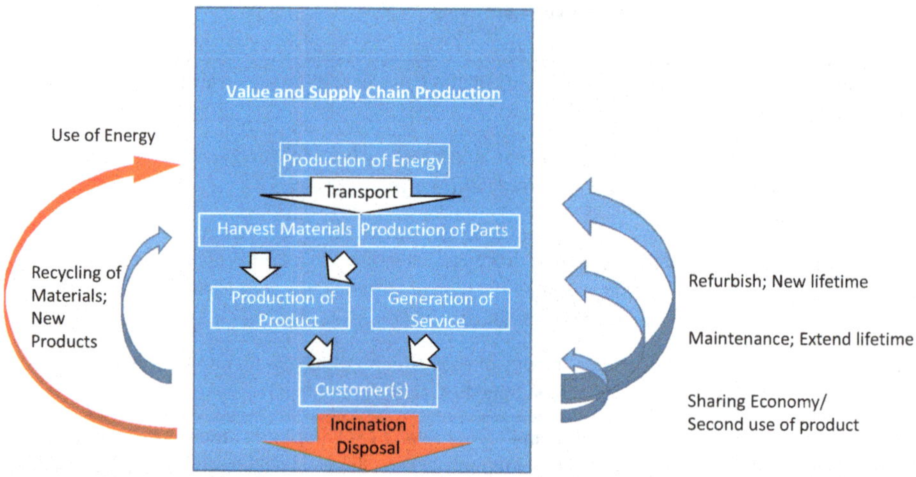

Figure 3.1: Material flows in the circular economy for a technical product according to the Ellen McArthur Foundation (circular economy diagram (ellenmacarthurfoundation.org)). The different cascades refer to the most efficient form of utilisation to the ones that go along with more consumption – which may be lesser preferred. At the end of all processes, waste leads to emissions and landfill.

It goes without saying that even if reuse is applied optimally, there must be a limitation for reuse of materials. Hence, in all cases, the best way to reduce the consumption of energy and materials is reduction of consumption.

Circular economy requires thinking about the system as a whole in the design process of the product. As can be seen in the examples below, all actors (companies, people and organisations) are part of one network in which the actions of one actor have an impact on other actors. In a circular economy, this is taken into account by considering both short- and long-term consequences of a decision and considering the impact on the entire value chain with the goal of a more resilient overall system. It is important to think in this way as, as a rule of thumb, >80% of all product-related environmental impacts can be influenced during the product's design phase. Failure to consider these aspects leads to waste – less than 2% of the material entering an economy remains in it for more than 6 months, the rest becomes waste sooner.

Sustainable product design, focused on resource efficiency, is critical to the shift to a circular economy because it aims to minimise consumption by recovering and reusing resources. Circular economy for a product involves not only environmental management, procurement, research and development and innovation processes in a sustainability strategy, but also business partners and customers and takes into account that

- recycled material tends to increase the concentration of hazardous substances in the process over time, requiring each actor to take this tendency into account and limit the use of such substances;
- the suitability of recycled materials for their downstream use already in the design process;
- recycled materials are often poisoned by remains of previous products. Such as, for example, plastic containing residues of off-spec chemicals cannot be readily considered for subsequent use in all applications (e.g. as food packaging).

3.3.1 Natural capitalism

Lovins' theory of natural capitalism [15] develops a model in which resources feed an economy consisting of services provided by nature (called ecosystem services). The resulting form of capitalism, called "industrial capitalism", is seen as a system that does not act responsibly because it "does not fully comply with its own accounting principles. It liquidates its capital and calls it income. It neglects to assign any value to the largest stocks of capital it employs – the resources and living systems and the social and cultural systems that are the basis of human capital." These stocks of capital are in turn seen as the capital on which "natural capitalism" is built, as opposed to industrial capitalism. The authors refer to "natural capital . . . to the earth's resources and the ecological systems that provide vital services to society and all living things. These services are of immense economic value; some are literally priceless as there is no known substitute for them. Yet current business practices typically do not take into account the value of these goods – which increases with their scarcity" [15].

In this context, capitalism builds businesses to prevent natural capital from being degraded through wasteful use of resources. According to Lovins, natural capitalism involves four changes in business practices:
- Increasing the productivity of resources
- Shifting to biologically inspired production models and materials
- Shifting to a "service and flow" business model
- Reinvestment in capital

The business model of interest here, called the "service and flow" business model, intends to move the economy from a goods-based economy to solutions-based business models.

To give an example:
If a customer wants mobility, a company that wants to provide a solution should sell ways to get people or goods from A to B conveniently, instead of focusing on manufacturing cars or trucks. The paradigm shift from producing and selling goods to service flows to satisfy customer needs also implicitly implies that the producer retains ownership of the goods produced. This promotes a 'take-back' situation when the productive life of the tool is over and consequently remanufacturing and recycling is in the best interest of the owner of the product or tool. This redefines the way in which the real needs of customers are addressed and met, and therefore changes the nature of the relationship between customers and producers/service providers as expectations and legal relationships change [15].

3.3.2 Blue economy

From 2006 to 2008, ZERI (the Zero Emissions Research Institute and the Biomimicry Institute) worked on the nature's 100 best research project [16] from which blue economy (which, as an approach, has been branded) developed as a scheme for business models. Blue economy works in a principle-based way – the principles of blue economy regard emissions and waste as misdirected, locally available and, if properly chosen, re-generable resources [16] as a resource for products/production. In this sense, the waste from one product automatically forms the raw material for a new product, that is, a cascade that theoretically has no final step [17]. This cascade approach does not lead to a single business model, but to a multitude of examples that harvest waste from previously unconnected value chains with new value chains and means to satisfy customer needs in partly new business models. To promote the idea, the website www.blueeconomy.org was set up to show examples where this concept has been implemented.

3.3.3 Regenerative design

The concept of regenerative design for products and related businesses is essentially focused on buildings, but the mindset advocated can be extended to other product classes to some extent. Similar to the principles of the blue economy, regenerative design starts from the observation that nature only works with what is available on site. It follows that sustainable economy must be based on local resources (the approach also refers to culture and tradition as important factors for the design of products), taking into account the same local environment in which it is used and in which it is recycled at the end of its life. To some extent, the approach is in contrast to standards, as they focus on one-size-fits-all solutions, as opposed to local solutions. Products or services based on regenerative design contribute to local ecosystems that are self-renewing or replenishing, in the sense, that once materials and energy have gone into a product or process, little to no further input is needed to sustain it. Certain priorities must be accepted and understood if such a design process is to be realistic:
- the relationship with a defined place throughout its life cycle (a partnership with a place rather than merely taking from it),
- the objectives in terms of regenerative capacity and
- the way in which harmonisation can be achieved between the place and the people who live there.

In this view, regenerative design implies the inclusion of systems thinking beyond the technology in question; it must also include interdisciplinary collaboration and consider local dependence on natural capital.

As this definition evolves, the primary goal of green building continues to increase the efficiency of energy, water and material use while reducing local and global environmental impacts.

Regenerative enterprises use or work with [18]
- systemic design that fits into ecosystems while being productive for people,
- materials that are abundant,
- nature as a compass for innovation (biomimicry),
- a long-term framework that involves all partners to harvest synergies, accepting each other's value and building on it to create business models and
- locally available and stranded assets.

It is in the logic of such an approach that local solutions, even if successful, cannot simply be copied from one proximity to another – copying must be accompanied by adaptation that takes into account locally available resources and culture as a crucial constraint. Local markets will have their own specific requirements, and the local partners responsible for commercialisation will also each bring their own views and needs to the table. It follows that such a product will not conform to global norms, but will be regionally specific, that is, a variant adapted to local resources and the market.

Scaling a regenerative business is complex, as the localisation efforts involved in establishing a business model can be prohibitive to its growth. In this sense, it is not just about return on investment and cash flow, but a more holistic approach to benefits (involving the local ecosphere, including people, but without neglecting other species) and creating win–win–win situations (for producers, consumers and the environment). It is in the logic of such regenerative enterprises that they go through phases of evolution: first growth, then maturity, and then with the need to adapt again if environmental conditions should change so that the products can remain fit for purpose. Therefore, it is helpful for companies to have a diverse portfolio in order to play a role in the local ecosphere.

A regenerative design approach to agriculture has been called permaculture [19], a combination of the words permanent agriculture or permanent culture, and means an ecological design system to promote the design of human habitats and food production systems based on the relationships and processes found in ecological communities [20]. Much of the inspiration has been drawn from the relationships and adaptations of indigenous peoples to their ecosystems. By creating "man-made ecosystems", the permaculture concept was able to show how to support human needs while reducing dependence on environmentally destructive industrial practices. Permaculture was the first ecological design system to introduce the concept of the regenerative effect as the standard for ecological performance of the built environment – the generation of a surplus or abundance of energy and resources that can be reinvested to further develop living systems as an integrated whole. He defined a sustainable built environment as one that is resource efficient and has minimal or neutral environmental impacts. In support of this objective, Mollison [20] introduced an investment hierarchy (regenerative, generative and degenerative) as a framework for assessing the value of potential measures to build regenerative capacity in a system.

3.4 Criteria for evaluation and benchmarking for economic sustainability

Many companies review their business models from time to time more and more also in the light of whether they offer some ways to incorporate sustainability. For an overview of the related benchmarking methods for corporate sustainability, see [21]. In the following, some of these methods will be briefly discussed.

3.4.1 SDGs of the United Nations

In the most general sense, the so-called sustainable development goals (SDGs [23]) developed by the United Nations help. Although not the primary purpose of these goals, they provide a way for companies to identify and benchmark themselves or their products with what has been defined by the United Nations. The SDGs were adopted by 193 countries at the United Nations in September 2015. The 17 goals and 169 targets are intended to inspire action in areas of critical importance to humanity and the environment:

1. Eradicate poverty
2. Eliminate hunger
3. Establish good health and well-being
4. Provide quality education
5. Enforce gender equality
6. Improve clean water and sanitation facilities
7. Generate affordable and clean energy
8. Create decent work and economic growth
9. Promote industry, innovation and infrastructure
10. Reduce inequality
11. Mobilise sustainable cities and communities
12. Influence responsible consumption and production
13. Organise climate action
14. Develop life below water
15. Promote life on land
16. Ensure peace, justice and strong institutions
17. Build partnerships for the goals

According to the International Chamber of Commerce, these SDGs can be seen as "a new lens for all companies through which they can translate the world's needs and ambitions into business solutions" [24]. They thus become values with which responsible customers can identify.

3.4.2 Measures and criteria for the transition phase

However, it is difficult to identify practical every day targets from these benchmarks for short- or medium-term pathways that is to support an economy in transition from being a "carbon junkie" to one that seeks to consume materials and energy sustainably. According to Charter et al. [25], the most important of the factors to consider are:

- changes in consumer sensitivity to sustainable solutions,
- the economic or legal need to gain a competitive advantage through sustainable strategies,
- the finite nature of (local) resources and their increasing costs for products (and their use),
- the increasing number of legal standards relating to pollution and environmental protection in relation to products, their use and their environment and
- increasing stakeholder awareness of the need for greater sustainability responsibility.

The VDI Guideline 2243 [26] offers a cascade model to identify the need for action regarding sustainability for an existing or planned product. Overall and for intermediate steps, this guideline focuses on three questions:

- Can the energy and resource demand be reduced by modifying (parts/modules of) the product?
- Can the lifetime of the product or parts of the product be extended (compared to the existing product approaches)?
- Are there ways to reduce the throughput rate of materials in the product life cycle?

3.4.3 Resolve methodology

Another way to test business models is through the ReSOLVE methodology.

The model starts from the observation that business models can support the implementation of a circular economy through the reuse of materials. The ReSOLVE methodology framework [27] can serve as a guide for the optimal use of materials – but needs to be adapted to local conditions and for a specific product portfolio. ReSOLVE was developed with the aim of reducing municipal waste in Poland (Table 3.3).

In this approach, companies can align their activities according to the six principles – regenerate, share, optimise, loop, virtualise and exchange (see [27]).

3.4.4 Emission pricing

The above-mentioned measures provide tools that, in a certain sense, require a rethinking towards new criteria. One way of thinking that does not require a rethink is to include prices for "footprints", especially the CO_2 emission footprint, in an economic consideration. Pricing has the potential to make products that are associated with high emissions uneconomical. With a tool of this kind, the goal of reducing emissions by 55% by 2030 (in the EU, approx. 2–3% reduction per year must be achieved – globally, 7–8% per year would be required (as stated by the United Nations at the end of 2020)) can be strongly motivated and in "the language of today's business models". For this approach, the European CO_2 emissions trading system, for example, made the large emitting industries pay a price of 25 euros per tonne of CO_2 emitted. This has to be increased to 55 euros by 2025. Norway plans to increase its CO_2 price to almost 200 euros per tonne by 2030, which is about three times the current level. In addition, emission budgets are to be introduced for each year. Currently, Sweden still leads the international ranking with just under 140 euros per tonne. In Germany, the Federal Environment Agency has calculated a CO_2 price of 180 euros as cost-covering on a worldwide scale in 2020.

Table 3.3: The ReSOLVE method as a framework that can be used as a guide on the way to a circular economy.

No.	General principle	Area applying for municipal waste	Description examples (perspective of possible application of proposed actions to the Polish conditions with respect to municipal waste)
1 Regenerate	Use renewable energy sources, rebuild ecosystems and return biological resources to the biosphere	Energy, heat or process steam recovery	Installations for the thermal transformation of municipal waste with energy recovery
		Reclaiming, retaining and restoration of health of ecosystems returning recovered biological resources to the biosphere	Landfill remediation Use of selected municipal waste fractions (e.g. from urban greenery) for fertilising purposes
		Sharing the products with co-users	Cohousing – sharing of joint areas in flats or houses Clothes sharing, e.g. leasing or sharing of clothes, such as T-shirts and jeans
2 Share	Share products with other users, second-hand use, life extension through product design and maintenance	Reuse of products by keeping the product loop speed low and maximisation of the utilisation of products	Donation for free or exchange for another product/service (clothes swap, toy swap) sale/resale of used goods and second-hand products
		Increasing product/technology performance and efficiency	Implementation of the most optimal solutions possible in the waste recovery and disposal processes

(continued)

Table 3.3 (continued)

No.	General principle	Area applying for municipal waste	Description examples (perspective of possible application of proposed actions to the Polish conditions with respect to municipal waste)
3 Optimise	Increase efficiency and effectiveness	Removal of waste from production processes	Comprehensive management of all waste streams
		Keeping the components and materials closed	Creation of reuse points and repair points, and creation of food banks
		Remanufacturing products or components, and, as a last resort, recycling materials	Repair points, eco-design
4 Loop	Remanufacturing of products, processing of materials into secondary raw materials, decomposition of biomass and chemical recovery from organic waste	Recycling and recovery of raw materials from waste streams	Increasing the efficiency of selective collection at source, including municipal biodegradable waste, in order for easy application of recycling/recovery technologies
5 Virtualise	Direct and indirect dematerialisation of products (eBook, online shopping)	Buying and using the utility virtually	Introducing virtual solutions in everyday life to reduce the amount of generated waste (such as newspapers, books, alarm clocks, music and online shopping)
6 Exchange	Replacing old with new technologies, products or materials	Replacing old with new	Replacement of household appliances and items (such as refrigerators, dishwashers, freezers) by items with a preferable energy class

3.4.5 Assessment at national levels – beyond GNP

There is no easy way to expand the instruments used today to evaluate economic success in such a way that sustainability is also brought into the equation. At present, economic policy pursues the goal of optimising the economic prosperity of the population. For this purpose, the gross national product (GNP or GDP) is used, which is seen as an agreed and measurable indicator of prosperity.

Nevertheless, one can argue about how GNP is measured, what limitations there are, what is measured or not, what additional information might be considered necessary for the generation of other and possibly more relevant indicators of progress. If policymakers consider GNP as the only relevant parameter, they may overlook other aspects that contribute to well-being (or well being) in non-monetary ways. Therefore, it may be useful to include other parameters that measure both economic performance and social progress.

This is important in the context of sustainability as the exploitation of nature is driven by the desire for wealth and profit by individuals and corporations. The goal of economic policy is prosperity – measured in macroeconomic terms such as GNP. Consequently, when policies are guided only by GNP as a measure of success, the broader well-being of citizens is not adequately considered. This is not to say that GNP is not important, but that there is evidence that an expanded measurement system more realistically reflects what needs to be measured. The question arises because many people already consider the state of nature as something that is crucial to their well-being.

At the beginning of the 1980s and in the aftermath of the 1968 movement, the fundamental concepts of economics were called into question. It became clear that the existing development paradigm of the 1970s, which measured the acceleration and deceleration of GNP, did not imply the increasing/decreasing happiness of individuals. Some of the proposed approaches seemed naïve, while others opened up new insights.

To address the need to accelerate happiness in lieu of monetary values and to challenge the conventional, materialistic notions of humanity, the fourth King of Bhutan (King Jigme Singye Wangchuck) spoke about Gross National Happiness (GNH). He made it clear that money does not equal happiness and that, in his opinion, material progress is not the most important contributor to well-being and therefore cannot be considered as a driving force of economic activities [32]. More specifically, the GNH is based on four pillars and nine dimensions.

The four pillars describe the

- the promotion of sustainable development,
- the preservation and promotion of cultural values,
- the preservation of the natural environment and
- the establishment of good governance.

The nine dimensions are given in Table 3.4.

Table 3.4: Dimensions of the GNP approach implemented in Bhutan according to [28].

- Education
- Psychological well-being
- Health
- Time use
- Cultural diversity and resilience
- Good governance
- Community vitality
- Ecological diversity and resilience
- Standard of living

If the above points are addressed and become the focus of a country's values, then the GNH should be able to describe a happier and thus more desirable direction of development.

A similar approach is developed in France at the end of the first decade of the new millennium.

In 2008, French President Nicholas Sarkozy, asked Joseph Stiglitz (Chair of the Commission), Amartya Sen (Advisor) and Jean Paul Fitoussi (Coordinator) to create a commission, later called "The Commission on the Measurement of Economic Performance and Social Progress", to scientifically investigate the need for statistical information on the economy and society. The aim of the Commission was to highlight the limitations of GDP as an indicator, to assess the feasibility of alternative measurement tools and to discuss how statistical information could be presented in an appropriate manner. The resulting report on measuring economic performance and social progress was published under the title "Mis-measuring our lives" [29]. In essence, the report concludes that the statistics commonly used do not capture some phenomena that have an increasing impact on citizens' well-being. The Stiglitz Commission, like the then King of Bhutan, concluded that this was the case and made a number of recommendations that, if considered in parallel with GDP, would better determine the national state of well-being and thus provide a more reliable indicator for any necessary government action (Table 3.5).

With these extensions, it should be possible to expand GNP and the corresponding set of instruments for steering economic policy in such a way that not only monetary wealth and prosperity are taken into account but also other criteria such as sustainability.

It is implicit to all the approaches discussed that consumption is not the only yardstick for analysing the well-being of nations, people and nature. This is important because it opens the way to an acceptance that consumption can actually decrease without compromising well-being. Implementing a circular economy should

Table 3.5: Proposed changes by the Stieglitz Commission to the GNP-based measurement of a nation's state of well-being [30].

1. When assessing material well-being, look at income and consumption rather than production.
2. Emphasise the household perspective.
3. Consider income and consumption together with wealth.
4. Place greater emphasis on the distribution of income, consumption and wealth.
5. Extend income measurement to non-market activities.
6. Quality of life depends on people's objective conditions and capabilities. Steps should be taken to improve people's measures of health, education, personal activities and environmental conditions. In particular, significant efforts should be made to develop and implement robust, reliable measures of social connections, political voice and insecurity that have been shown to predict life satisfaction.
7. Indicators of quality of life in all dimensions covered should comprehensively assess inequalities.
8. Surveys should be designed to capture the interrelationships between different quality of life domains for each person, and this information should be used in designing interventions in different domains.
9. Statistical offices should provide the information needed to aggregate across different quality of life dimensions to enable the construction of different indices.
10. Measures of both objective and subjective well-being provide important information about people's quality of life. Statistical offices should include questions to capture people's life assessments and priorities in their own surveys.
11. Assessing sustainability requires a well-defined dashboard of indicators. The components of this dashboard should be characterised by being interpretable as variations of some underlying "stock". A monetary sustainability index has its place in such an indicator system, but should remain essentially focused on the economic aspects of sustainability, given the current state of the art.
12. The environmental aspects of sustainability deserve separate consideration based on a well-selected set of physical indicators. In particular, a clear indicator of proximity to hazardous environmental damage is needed (e.g. related to climate change or depletion of fish stocks).

reduce waste and helps prevent ecological disaster – but it is only part of the solution. It is unlikely that the circular economy will ever be 100% closed, and therefore reducing consumption also becomes an important part of the equation.

For an overview of the topic, reading *An Introduction to Ecological Economics* is recommended [31].

References

[1] Ovans, A., What Is a Business Model?, Harward Business Review, 23. January 2015, https://hbr.org/2015/01/what-is-a-business-model

[2] Zott, C., Raphael, A.M.I.T., Lorenzo, M.A.S.S.A., 2011. The Business Model: Recent Developments and Future Research [online]. Journal of Management, 37(4),1019–1042. ISSN 0149-2063.

[3] Schaltegger, S., Hansen, E., Lüdeke-Freund, F. (2016): Geschäftsmodelle für Nachhaltigkeit: Origins, Present Research, and Future Avenues (Editorial), Organisation & Umwelt, Vol. 29, No. 1, 3–10.
[4] https://bstrategyhub.com/50-types-of-business-models-the-best-examples-of-companies-using-it/, 50 Types of Business Models (2020).
[5] Business model – Wikipedia, https://en.wikipedia.org/wiki/Business_model
[6] James Richardson: The business model: an integrative framework for strategy execution. In: Strategischer Wandel. Band 17, Nr. 5-6, 2008, ISSN 1086-1718, S. 133–144, doi:10.1002/jsc.821.
[7] Osterwalder, A., Pigneur, Y., Tucci, C.L.: Clarifying Business Models: Ursprünge, Gegenwart und Zukunft des Konzepts. In: Mitteilungen Der Gesellschaft Für Wirtschaftsinformatik. Band 15. Association for Information Systems, 2005
[8] Krumeich, J., Burkhart, T., Werth, D., Loos, P.. Towards a Component-based Description of Business Models: A State-of-the-Art Analysis. Americas Conference on Information Systems (AMCIS 2012) Proceedings. Paper 19.
[9] http://aisel.aisnet.org/amcis2012/proceedings/EBusiness/19).
[10] Lüdeke-Freund, F., WORKING DEFINITIONS OF "SUSTAINABLE BUSINESS MODEL" & "BUSINESS MODEL FOR SUSTAINABILITY" June 9, 2014, https://sustainablebusinessmodel.org/2014/06/09/working-definitions-of-sustainable-business-model-business-model-for-sustainability/
[11] Sirgy, M.J., Danes, J.E. (1982), "Self-Image/Product-Image Congruence Models: Testing Selected Models", in NA – Advances in Consumer Research Volume 09, eds. Andrew Mitchell, Ann Arbor, MI: Association for Consumer Research, Pages: 556–561.
[12] vgl. Osterwalder, A., Pigneur, Y., Tucci, C.L., 2005. Klärung von Geschäftsmodellen: Origins, Present and Future of the Concept. Mitteilungen Der Gesellschaft Für Wirtschaftsinformatik 16, Artikel 1.
[13] Bocken, N.M.P., de PAUW, I., Bakker, C., Bram, VAN DER GRINTEN, 2016. Produktdesign und Geschäftsmodellstrategien für eine Kreislaufwirtschaft. In: Journal of Industrial and Production Engineering [Online]. 26.04.2016 [Zugriff am 4.11.2020]. Nr. 33/5, S. 308–320
[14] Wirtz, B. (2001): Electronic Business, Wiesbaden 2001, S. 151
[15] Natural Capitalism: Creating the Next Industrial Revolution, gemeinsam verfasst (mit Paul Hawken) von den Co-CEOs des Rocky Mountain Institute, Amory und Hunter Lovins. http://www.naturalcapitalism.org
[16] Pauli, G.A., The Blue Economy: 10 Years, 100 Innovations, 100 Million Jobs, Paradigm Publications. April 2010, https://books.google.de/books?id=aJ3HZD1H7ZsC&printsec=frontcover&hl=de#v=onepage&q&f=false
[17] https://web.archive.org/web/20141101151622/http://www.blueeconomy.de/page/dieprinzipien
[18] https://medium.com/swlh/how-to-scale-a-regenerative-business-model-601a4d972cc8
[19] Mollison, B., Permaculture: a designers' manual. Tagari Publications, Australien, 1988.
[20] Mang, P., Reed, B., Regenerative_Design_and_Development, Regenerative Development and Design, Regenesis Group and Story of Place Institute, 2015, www.regenesisgroup.com/wp-content/uploads/2015/02/Encylopedia_Sustainability_Science_Ch303.pdf
[21] Mollison, B. (1988) Permaculture: a designers' manual. Tagari Publications, Australien.
[22] Rudolph, C., 2018. Geschäftsmodell Circular Economy: Gegenwart und Zukunft der (erweiterten) Kreislaufwirtschaft. In: Bungard, P., Hrsg. CSR Und Geschäftsmodelle. Auf Dem Weg Zum Zeitgemäßen Wirtschaften. Berlin: Springer Gabler, S. 123–137. ISBN 978-3-662-52882-2.
[23] https://sdgs.un.org/goals

[24] https://worldtop20.org.

[25] Charter, M., Peattie, K., Ottman, J., Polonsky, M.J., 2006. Marketing und Nachhaltigkeit. [Online] Verfügbar unter: www.cfsd.org.uk/smart-know-net/links/smart-know-net.pdf

[26] VDI-Richtlinie 2243 (Eine Richtlinie zur Produktentwicklung bzw. Produktänderung; Recyclingorientierte Produktentwicklung, VDI-Gesellschaft Produkt- und Prozessgestaltung, Frankfurt, 2002–07.

[27] Regenerate, Share, Optimize, Loop, Virtualize, and Exchange; Marzena Smol, Joanna Duda, Agnieszka Czaplicka-Kotas, Dominika Szołdrowska, Transformation towards Circular Economy (CE) in Municipal Waste Management System: Model Solutions for Poland, June 2020, Sustainability 12 (11):4561; DOI: 10.3390/su12114561; www.ResearchGate.net

[28] Wolfgang Sachs, Die Archäologie der Entwicklungsidee, https://www.burmalibrary.org/docs14/The_Archaeology_of_the_Development_Idea.pdf; abgerufen am 11. April 2020.

[29] Stiglitz, J., Sen, A., Fitoussi, J.-P.: Mismeasuring Our Lives: Why GDP Doesn't Add Up. The New Press, New York 2010, ISBN 978-1-59558-519-6.

[30] Empfehlungen des Stiglitz-Sen-Fitoussi-Reports; Marie C LERC, Mathilde G AINI, Didier B LANCHET, INSTITUT NATIONAL DE LA STATISTIQUE ET DES ÉTUDES É CONOMIQUES, Série des documents de travail de la Direction des Études et Synthèses Économiques, AVRIL 2011.

[31] An Introduction to Ecological Economics; Robert Costanza, John H Cumberland, Herman Daly, Robert Goodland, Richard B Norgaard, Ida Kubiszewski, Carol Franco, Boca Raton, CRC Press 2015, ISBN 9781566706841).

[32] Bhutan's Gross National Happiness Index, Oxford poverty and Human Development Initiative; https://ophi.org.uk/policy/national-policy/gross-national-happiness-index/ Retrieved April 11, 2020

Chapter 4
Life cycle assessment (LCA)

4.1 Introduction

As described, the life cycle of a product involves a series of value-adding steps, from the extraction of natural resources to the end of the product's life. In a circular economy system, the end of life of the product is reconnected with its production by reusing the already extracted resources contained in used products. This circular economy system is particularly suitable for certain types of products where there is the possibility of producing new products from used old products (established, e.g. for glass and metals). Life cycle assessment (LCA) is defined as an objective process for evaluating the environmental impacts associated with a product [1], process or activity [2] by identifying the energy and materials consumed and the waste released into the environment, and for assessing and implementing opportunities for environmental improvement. LCA aims to quantify the environmental impact over the product's life cycle [4,5,6].

Most sources trace the origins of the term "life cycle assessment" for an ecological balance to a study by Harold Smith, project manager for the Douglas Point Nuclear Generating Station, Canada. At the World Energy Conference in 1963, Smith reported on his calculation of the cumulative energy demand for the production of chemical intermediates and products. Later, several global modelling studies were conducted, leading to further dissemination of these techniques for resource analysis (the precursors of LCA). In particular, the publication of *The National Academy of Science's Resources and Man* (1969), Dennis Meadows' book *The Limits to Growth* (1972) and the Club of Rome's document *A Blueprint for Survival* (1972) (see Chapter 1) led to predictions of the impact of changing world population and the expansion of industrial processes on the demand for finite raw materials and energy resources. Meanwhile, the ISO 14000 standard for LCA was developed in the 1990s. In 2006, ISO updated the 14040 and 14044 standards and in 2008, BSI [7] published the "Specification for the assessment of life cycle greenhouse gas emissions from goods and services" after consultation. Conducting an LCA can be resource and time intensive. Depending on how thorough an LCA is to be conducted by the user, data collection can be problematic and the availability of data can greatly affect the accuracy of the final results – see also EPA 101 (2006).

4.2 Steps to take in the standardised LCA process

1. Goal and scope definition

This phase is primarily concerned with defining system boundaries and relationships to overlapping product life cycle. Within the boundaries, there is the definition of what is being considered/what is to be understood (Table 4.1) and what the environmental

https://doi.org/10.1515/9783110767308-004

impacts are and how they relate to it. When it comes to comparisons, there is a clear definition of what is being compared.

2. Life cycle inventory (LCI) – what is in the product?
An LCI is a kind of bookkeeping. It involves the compilation of lists, which in turn contain all the components of the products that are used or needed during the life cycle of a product, are created and fall within the defined system boundary (see Table 4.1).

As seen in the example below, it includes three steps:

A supply chain that shows
- Raw materials
- Manufacturing processes
- Transports
- Uses
- Waste management

The data for
- Material input
- Products and by-products
- Solid waste, air and water emissions

The calculation of the relevant quantities
- For example, in relation to the functional unit

The ultimate goal of life cycle analysis is to collect the relevant data for a product for all phases of its life cycle and to correctly allocate these data to these phases (including the relevant supply and waste chains).

3. Life cycle impact assessment (LCIA) – impacts.
Classification is about assigning specific environmental impacts to each component of the LCA. This is where the decisions made in the scope and target phase about which environmental impact categories are of interest come into play.

See Table 4.2 [8].

4. From values to weights
Often, once impact categories have been identified, the respective parameter can be normalised and weighted using directly comparable impact indicators. The advantage of variable weighting approaches is that they can be adapted to the goals and values of an organisation. There are some standards for sustainability accounting and reporting that focus almost exclusively on GHG emissions, so it makes sense for organisations using them to place almost all of their weight on this subset of impact factors (e.g. [7]).

Table 4.1: Data to be considered in an LCA of a material product.

Metrics/attributes	Parameter
Consumption metrics	Fossil fuel Water Biotic resource Mineral
Emission metrics	Greenhouse gas Clean production: human impacts Clean production: aquatic toxicity Eutrophication
Packaging attributes	Content (recycled or virgin) Sourcing Solid waste Material health

Table 4.2: Frequently used impact categories according to [8].

Impact category	Scale	Examples of LCI data (i.e. classification)	Common possible characterisation factor	Description of characterisation factor
Global warming	Global	Carbon dioxide (CO_2) Nitrogen dioxide (NO_2) Methane (CH_4) Chlorofluorocarbons (CFCs) Hydrochlorofluorocarbons (HCFCs) Methyl bromide (CH_3Br)	Global warming potential	Converts LCI data to carbon dioxide (CO_2) equivalents Note: global warming potentials can be 50, 100 or 500 year potentials.
Stratospheric ozone depletion	Global	Chlorofluorocarbons (CFCs) Hydrochlorofluorocarbons (HCFCs) Halons Methyl bromide (CH_3Br)	Ozone depleting potential	Converts LCI data to trichlorofluoromethane (CFC-11) equivalents.
Acidification	Regional Local	Sulphur oxides (SO_x) Nitrogen oxides (NO_x) Hydrochloric acid (HCl) Hydrofluoric acid (HF) Ammonia (NH_4)	Acidification potential	Converts LCI data to hydrogen (H^+) ion equivalents.

Table 4.2 (continued)

Impact category	Scale	Examples of LCI data (i.e. classification)	Common possible characterisation factor	Description of characterisation factor
Eutrophication	Local	Phosphate (PO_4) Nitrogen oxide (NO) Nitrogen dioxide (NO_2) Nitrates ammonia (NH_4)	Eutrophication potential	Converts LCI data to phosphate (PO_4) equivalents.
Photochemical smog	Local	Non-methane hydrocarbon (NMHC)	Photochemical oxidant creation potential	Converts LCI data to ethane (C_2H_6) equivalents.
Terrestrial toxicity	Local	Toxic chemicals with a reported lethal concentration to rodents	LC50	Converts LC50 data to equivalents; uses multimedia modelling, exposure pathways.
Aquatic toxicity	Local	Toxic chemicals with a reported lethal concentration to fish	LC50	Converts LC50 data to equivalents; uses multimedia modelling, exposure pathways.
Human health	Global Regional Local	Total releases to air, water and soil	LC50	Converts LC50 data to equivalents; uses multimedia modelling, exposure pathways.
Resource depletion	Global Regional Local	Quantity of minerals used Quantity of fossil fuels used	Resource depletion potential	Converts LCI data to a ratio of quantity of resource used versus quantity of resource left in reserve.
Land use	Global Regional Local	Quantity disposed of in a landfill or other land modifications	Land availability	Converts mass of solid waste into volume using an estimated density.
Water use	Regional Local	Water used or consumed	Water shortage potential	Converts LCI data to a ratio of quantity of water used versus quantity of resource left in reserve.

5. Evaluation

This phase goes hand in hand with the review, quantification and evaluation of the input provided, used and resulting from the LCI and LCIA phases. The standard that covers LCA, ISO 14044, calls for:

- Analysing results, drawing conclusions, explaining limitations and making recommendations based on the findings of the previous phases of the LCA, and report the results of the LCA in a transparent manner.
- Providing an easily understandable, complete and consistent presentation of the results of an LCA study in accordance with the objective and scope of the study.

To achieve this, the ISO 14044 standard specifies that the interpretation should include at least three main elements:

- Identification of significant issues based on the LCI and LCIA. Significant phases or contributors per phase or components, anomalies – the easy targets to achieve.
- Cross-check for completeness, sensitivity and reliability areas covered.
- Cross-check for consistency of assumptions, data, characterisation factors, etc. used.
- Conclusions, recommendations and reporting.

In the final analysis, the goal is to optimise the economic aspect of a company's value creation considering its consumption characteristics. This is no different with a closed-loop supply chain – except that here the environmental aspect becomes part of the equation. The analysis ultimately aims to reduce emissions, consumption of non-renewable/scarce resources and energy consumption. Economic benefits arise in terms of emissions mainly from the reduction of emission costs, transport, processing and utilisation costs and the minimisation of the use of materials to be retained. The implementation of a closed-loop supply chain thus enables both ecological and economic value creation.

As can be seen in the following example, the definition of the product and e.g. consumables and expendable goods or the service that are inevitably used during the life phases of the product form the system boundaries in which the environmental impacts are checked. The data collection provides the basis for an estimation of resource consumption (e.g. water, energy and materials), waste and emissions (in CO_2 equivalents) in the different life phases. The evaluation leads to assigning a relative value to different concepts of eco-design and enables an economic and sustainable assessment of the proposed designs.

Often the use of the product by customers is only a small part of the life of the product (think of the short time water spends in a household compared to the long journey water takes during manufacture for use, transport, cleaning and return). The value of careful eco-design to the economy of a company or society can be significant – especially when all life stages are not just about use, but also about minimising material and energy consumption for use, as well as enabling access to all goods for later reuse – whether in the same or different life cycles.

The generation of an LCA is labour and data intensive. For some applications, it is advantageous to use the life cycle method but without going through a full LCA.

Table 4.3: Techniques for streamlining an LCA.

- Adjusting the system boundaries (reducing the number of processes or life cycles involved).
- Limiting the types and number of resources, pollutants or environmental impact categories.
- Eliminating steps (e.g. mining of raw materials in favour of other sources of raw materials)
- Using qualitative information or estimates to supplement or replace numerical data.

Simplified methods are generally referred to as "streamlined LCA". Examples of the techniques used in streamlining an LCA are given in Table 4.3.
Some examples of applied LCAs can be seen on the following company websites.

It is very important to note that LCAs, no matter how carefully they have been compiled, analysed, evaluated and measured, are never answers in themselves. They need to be interpreted, which in turn requires judgement. The data sources, assumptions and all other relevant information must be transparent to decision-makers so that they can understand the full context of the LCA results. Deciding between different design options is not as simple as just comparing LCIA figures, whether one-dimensional or multidimensional, weighted or not. LCIA results can be a source of insight, but they do not stand alone for making product development decisions. Engineers need to see them in the context of the other attributes they are trying to optimise, including cost, manufacturability, performance and more. In addition, there are factors that guide product development decisions that are not covered by LCAs, including social impact and acceptability, pricing, political agendas and regulations.

References

[1] European Commission, Product Umweltfußabdruck Category Rules Guidance, https://ec.europa.eu/environment/eussd/smgp/pdf/PEFCR_guidance_v6.3.pdf, Retrieved April 11, 2020.

[2] European Commission, Organisation Umweltfußabdruck (OEF) Guide https://ec.europa.eu/environment/eussd/pdf/footprint/OEF%20Guide_final_July%202012_clean%20version.pdf", 2012, Retrieved April 11, 2020.

[3] DIRECTIVE 2009/125/EC OF THE EUROPEAN PARLIAMENT AND OF THE COUNCIL of 21 October 2009 establishing a framework for the setting of Eco-Design requirements for energy-related products https://eur-lex.europa.eu/legal-content/EN/ALL/?uri=CELEX%3A32009L0125, Retrieved April 11, 2020.

[4] Hoekstra, A., Wiedmann, T., Humanity's unsustainable environmental footprint, http://www.ayhoekstra.nl/pubs/Hoekstra-Wiedmann-2014-EnvironmentalFootprint.pdf, Retrieved April 11, 2020.
[5] Rees, W.E. (1992). Environmental Footprints and appropriated carrying capacity: what urban economics leaves out. Environment and Urbanization, 4(2),121–130. https://doi.org/10.1177/095624789200400212, Retrieved April 11, 2020.
[6] EU Commission, European Platform on Life Cycle Assessment (LCA), https://ec.europa.eu/environment/ipp/lca.htm, Retrieved April 11, 2020.
[7] ISO 14040 ff.
[8] A technique for assessing the environmental aspects associated with a product over its life cycle." Goedkoop et al. (2008).

Chapter 5
Environmental footprints and their calculation

When phrased in a commonly understandable and accessible way, footprints can help producers and users to better understand the resource and energy consumption going along with a product. Areas of high consumption or significant energy loss point to opportunities to improve efficiency by implementing resource management best practices, upgrading systems or developing new technologies. Footprints provide a benchmark to assess the benefits of efficiency improvements and to prioritise the analysis of opportunities.

5.1 CO_2 footprints

As mentioned greenhouse gas (GHG) emissions must be limited to prevent a rise in average atmospheric temperatures. To this end, most industrialised countries have introduced measures to limit such emissions as well as developed implementable measures to reduce them. To enforce this, an annual percentage reduction is often mandated by law. In order to develop appropriate reduction strategies in a company, it is first necessary to know the GHG emissions in a defined period in numerical terms – especially their main contributions. Consequently, methods are needed to assess the emissions and develop emission factor-related reduction strategies.

In the following, the methodology for assessing Greenhouse Gas Protocol (GHP) Scope 1, 2 and 3 emissions is used, and examples of major contributors to emissions are given. In connection with emissions and the increasing likelihood of legal proceedings, penalties and negative publicity, there are sometimes serious risks for each individual company – such risks are presented, structured and provided with examples.

5.1.1 GHG emission footprint and CO_2 footprint

GHGs are a class of substances known to cause an increase in the Earth's temperature in the atmosphere. The main GHGs are carbon dioxide, methane, nitrous oxide, sulphur oxide and the fluorinated gases (including hydrofluorocarbons, perfluorocarbons, sulphur hexafluoride and nitrogen trifluoride). Carbon dioxide accounts for 64.3% of GHGs. In order to work with only one figure and to scale the effect of individual gases on the climate, their contribution is converted into the mass of CO_2 which, when emitted into the atmosphere, has a comparable effect. This figure is called the "CO_2 footprint" or "carbon footprint" [1] – both terms are used synonymously in the following. The value is often given as a mass equivalent.

https://doi.org/10.1515/9783110767308-005

It is worth noting that, once emitted, some of these gases are known to remain in the atmosphere for hundreds of years, during which time they have a climate-relevant effect.

There are many laws limiting GHG emissions, e.g. the Climate Change Act [2]. In autumn 2020, the EU decided to make annual emission levels with a reduction target of "at least 55%" by 2030 legally binding for all EU member states [3]. The GHG as an emission indicator is based on the National Inventory Submissions 2019 [4]. A CO_2 or carbon footprint (the terms are used interchangeably) caused by a person, an event, an organisation, a service, an industry segment or a product summarises GHG emissions. The term was popularised by an advertising campaign by the oil and gas company BP [22] in an attempt to shift public attention away from limiting fossil fuel companies' activities and towards individual responsibility for solving climate change [5].

The carbon footprint is not an abstract number; it must be expressed in tonnes per a given reference – very often per year. The term refers to a concrete situation that sets the boundary conditions for what (not) to consider. Often and among other footprints, the carbon footprint is assessed as part of a life cycle assessment (LCA; see ISO 140040 for products).

5.1.2 GHP Scope 1, 2 and 3 emissions

In order to structure the carbon footprint and to facilitate analysis and conclusions for policies and related measures, CO_2 emissions need to be structured. Each industry has its own assessment and modelling techniques, allocation processes, mitigation methods and labelling strategies for its carbon emissions [1].

To allocate risks and define appropriate stakeholder strategies, it is helpful to structure emission contributions according to GHG Scope 1, 2 and 3 emissions using the GHP guidelines. To illustrate this approach, it is first necessary to look at the GHP [2]. The GHP defines the basic principles of relevance, completeness, consistency, transparency and accuracy and is based on financial accounting. In addition, there are efforts by the standardisation institutions DIN and ISO to establish corresponding standards.

The GHP defines the so-called scopes to enable a separation of internal (and easy to assess) emissions – Scopes 1 and 2 – from emissions caused by incoming and outgoing processes – Scope 3 – which are often very difficult to assess (Table 5.1).

Under practical conditions, it is not always obvious to decide which step in the value chain belongs, where analysis and the resulting strategies require expertise.

Table 5.1: Scopes.

Scopes
Scope 1: direct emissions – internal Scope 1 emissions refer to emissions that enter the atmosphere as a direct result of activities. These are all direct emissions resulting from the activities of a company or a resource controlled by the company, as well as from immobile sources. They include on-site fuel combustion, such as gas boilers and air conditioning leakage. As indicated, they may relate to a specific company, product or sector within a company.
Scope 2: indirect emissions Scope 2 emissions refer to emissions released into the atmosphere from the generation of purchased energy by a utility. These are indirect emissions from electricity purchased and consumed by the organisation, as well as from the consumption of purchased electricity, steam, heating and cooling, including mobile consumption, which occur during the production of the energy.
Scope 3: indirect emissions – external Scope 3 includes energy consumed upstream and downstream in the value chain, including in transmission and distribution [22]. – If not included in the other scopes – occurring in the value chain upstream and downstream of the reporting company or product. In other words, emissions are associated with the company's operations. According to the European Commission [2], scope 3 emissions are divided into 15 categories. Upstream activities include various categories such as business travel, employee commuting, waste generation and disposal, processes upstream logistics and capital goods [23]. Downstream activities include investments, franchises, leased facilities, downstream logistics, use of sold products, recycling, preparation for reuse, waste and disposal. These need to be assessed in the specific case with the relevant selection of criteria. For this, careful analysis of the given situation is crucial.

To illustrate this with an example:
When producing a printed product, the carbon footprint of the inbound value chain – especially the paper – and the production of the printed product must be taken into account. The situation is less clear when looking at the outbound value chain and the recycling situation: In the case of paper, the responsibility for

- collecting the printed products for waste papers

  or
- the print product for which the recycled paper is used.

At first glance, this differentiation seems academic, but it can be of serious financial significance for a newspaper, for example, if a CO_2 balance sheet has to be drawn up and disposal has to be taken into account as a cost factor:

As a rule of thumb, about 10–20% of the mass of paper entering a recycling chain at the waste paper bin is lost during the recycling process. This fraction is usually incinerated after it has been separated from the material to be processed. When this mass is burnt, it leads to emissions. Assuming that 1 tonne of paper corresponds to an emission equivalent of 1.15 tonnes of CO_2, this mass must be attributed to one of the responsible products and paid for. As shown below in the context of the risk analysis, the price per tonne of CO_2 footprint varies between €30 and €180 (in 2019 and in Sweden €140/tonne CO_2 footprint).

There is a reasonable discussion about the extent to which burning wood is CO_2-neutral or not: while the emissions are obviously attributable to combustion, the counter-argument is that the CO_2 emitted in combustion was previously captured from the atmosphere, but is only released in equilibrium through combustion. And if at least as many trees are planted as were cut down, the carbon removed and reintroduced into the atmosphere balance each other out in the end [3].

5.1.3 Emissions

The following examples reflect actual results originating of footprint analysis that were determined in two different companies. The aim of the study was to better understand and reduce CO_2 footprints in both companies, in particular to prepare for possible emission-related saving targets and costs (Table 5.2).

Here, the presentation of the categorisation of emissions follows the scopes of the G GP (see also emission categories under 25):

While scope 1 includes all direct emissions, i.e. emissions generated by combustion in a company's own facilities, scope 2 describes all emissions that are generated by a company itself.

Scope 2 describes all emissions associated with purchased energy (e.g. electricity and district heating).

Scope 3, in turn, includes indirect GHG emissions (e.g. from business travel or purchased goods and services).

For comparison, the carbon footprint of a company producing plastic parts in the Middle East is given in Table 5.3.

Table 5.2: CO_2 footprint of a newspaper printing company (rounded figures) producing mainly for the national market.

Category	Absolute CO_2 footprint in t CO_2 eq. p.a. (rounded)	Remarks
Newspaper		
Scope 1 (~0.5%)		
Stationary combustion engines	130	Mainly gas Diesel for power emergency negligible
	Emissions fluorinated gases	

Table 5.2 (continued)

Category	Absolute CO_2 footprint in t CO_2 eq. p.a. (rounded)		Remarks
Scope 2 (~2.8%)			
Indirect emissions	Purchased electricity	1,800	
Scope 3 (inbound, 81%)			
Purchased products and services	Paper	20,000	
	Offset ink	1,800	
	Liquids	–	
	Plates	400	Ration raw versus recycled material unknown
	Water	10	
	Tools and machines Press	350	Annual consumption assuming a predefined lifetime of the presses – not knowing whether this lifetime can be achieved
	Facility management	20	
	Loss caused by electricity transportation	150	
	Inbound transport – paper	1,200	
	Inbound transport – goods	10	Estimation assuming an average distance of transport of 400 km
	Non-print-related waste Not assessed		No records available
	Business travel	2	
	Commute	600	
Scope 3 (outbound, 16%)			
Outbound logistics		150	Assuming that readers are located within an average circle of 150 km
Total p.a.	~25,000 tonnes of CO_2		

Table 5.3: Rounded carbon footprint of a plastic parts production facility.

Category	Absolute CO_2 footprint in tonnes of CO_2 eq. p.a. (rounded)		Remarks
Scope 1 (~ 0.5%)			
Stationary combustion	Mainly gas	10	
	Emission fluorinated gases R134a	0.2	
	Process-related use Propane gas	10	
Scope 2 (~41%)			
Indirect emissions	Electricity	1,510	
	Petrol	30	
Scope 3 (inbound, ~43%)			
Purchased products and services	Liquid (solvents)	1	
	Raw materials	1,060	
	Water	240	
	Tools and machines Not assessed	–	Tools and machines partly aged and still in use. No documentation on system footprints or lifetime available.
	Facility management	20	
	General transport	170	
	General waste	45	
	Business travel	15	42 intern. flights
	Commute	35	
Scope 3 (outbound, ~16%)			
Outbound logistics		600	
Total rounded	~ 3,760 tonnes of CO_2 p.a.		

Discussion of the examples

In both examples, four to five categories contribute the largest share of emissions. If the goal of the assessment is to reduce emissions or the costs associated with them, it is clear that spending too much effort to capture the other emissions with a high degree of accuracy will not lead to action plans that contribute to a significant reduction in

CO_2 emissions. Therefore, when looking for strategies, it is advisable to look for "the big numbers" and ways to reduce the respective consumption.

When comparing the two examples, the absolute numbers offer the first visible difference. However, this size only reflects the size of the companies one with about 40 employees and the other with about 400.

The next difference is not merely product-related: The electricity produced (in this case mostly wind- and water-powered) has a much smaller carbon footprint than the diesel-powered generators used to produce electricity at the other production site.

In both cases, inbound logistics contribute significantly to the carbon footprint, while outbound logistics are small for the printing company under consideration, but large for the other producer. The difference is due to the business model: Both companies work under tight time constraints, but while one is located close to its customers, the other is located several 1,000 km away from the customer. Both require transport of the product to the customer, but given the tight delivery deadlines for the parts-producing company, it is necessary to use fast ships, which in turn are characterised by very high emissions.

The access to raw materials for both products also differs significantly: plastic is easy to obtain as a raw material for parts if large quantities are purchased. Paper, on the other hand, has become increasingly difficult to access in recent years – also due to the population's changed environmental awareness. This in turn reflects the distances that have to be travelled for raw materials.

As discussed above from a business perspective, the questions are:
- How large is the product/company-related carbon footprint?
- What are strategies to reduce it so that the company under consideration operates under legal conditions?
- What are the risks associated with emissions if the company acts/does not act?

It is not the intention here to discuss/suggest possible strategies from the figures presented or their relationship – the discussion of these questions is too company-specific and hardly offers any general added value – what is of interest, however, are the risks associated with non-sustainable action.

5.2 Economic risks linked to emissions

The German Federal Banking Supervisory Authority (BAFIN) has prepared a best practice document describing the associated risks [45]. In the following, the model presented in that source is used to address individual risks.

Of course, there are interdependencies between risks, especially between physical risks and transformation risks. For example, an energy transition that proceeds too

Table 5.4: Risks going along with the environmental impact of consumption.

Types of risk Risks can be organised by type – here they are grouped into physical, transitory (or transfer) and reputational risks.
Physical risks [45] Physical risks are the direct and indirect consequences of both individual extreme weather events and long-term changes in climatic and ecological conditions Specifically, these are: (i) Operational risk Example: A company or its suppliers are affected by an environmental disaster. (ii) Underwriting risk (relevant in banking and insurance business) Example: In non-life insurance, losses increase due to storms, floods, forest fires or hail. Likewise, losses in business interruption insurance may increase. A new type of intensity and/or frequency of such events is not (yet) adequately reflected in the technical provisions or in the measurement of the premium risk. (iii) Legal risk Example: Risks for environmental damage or partners in the supply chain who neglect sustainability run the risk of being held responsible for the consequences and having to change their business practices in this context. (This type of risk is not addressed in the following – even though, for example, changes in the supply chain could certainly cause problems for the plant under consideration.)
Transition risks (also called transfer risks) Transition risks are understood as risks for business models resulting from decarbonisation and the transition to zero- or low-carbon economic structures. The term was introduced by the UK Financial Services Authority in 2015 [45]. For example, the coal phase-out and carbon tax are leading to new challenges and investments across sectors. (i) Default risk Example: An expected or unexpected change in market sentiment (e.g. due to the pricing of expected regulatory measures) leads to devaluations or disruptions in the supply chain and contracts cannot be serviced as a result. (ii) Market (price) risk Example: Prices change unexpectedly due to changes in legislation or their consequences. (iii) Liquidity risk Example: After an environmental disaster, companies get into financial difficulties. (iv) Strategic risk Example: A company's strategic procurement partner loses its business because production processes become more expensive due to environmental regulations. (v) Stranded assets, i.e. assets (e.g. company shares, technical equipment or (raw material) inventories) whose earning power or market value unexpectedly drops drastically until they become worthless.
Reputational risks In addition to the effect of sustainability risks on the net assets, financial position and results of operations, there are also reputational risks. Here, the effect can arise from events and behaviour that have occurred or from business relationships with companies (external) or internal structures in the company with high sustainability risks.

Table 5.4 (continued)

Examples:
Internal/external – The environment or people are harmed due to inadequate internal guidelines on environmental standards. The case is reported in the media, with the companies being named. – Involvement in and own "greenwashing" strategies (at a company and its suppliers) can also pose a reputational risk if they become known and catch the eye. Worth mentioning here is compensation payments without own measures, cooperation with unchecked compensation payment partners or also false calculations with the aim of minimising one's own footprint.

slowly may cause more frequent and more severe damage and subsequently require a much more abrupt change in the economy, which increases transformation risks.

Financially presentable risks

In the following, the currently identifiable material risks are to be described and, as far as possible, assessed in monetary terms.

For a risk assessment, it is necessary to evaluate the probability of occurrence of the various risks or to form scenarios such as best or worst case specifically for a product or plant including the associated value chains and at least the most important/strategic partners. A detailed analysis including scenario can only be done company/product specific (see ISO 3100, Risk Management). In Table 5.5, therefore, only some of the risks are named and briefly described as well as generally assessed and the analysis only provides best- and worst-case scenarios for quantifiable risks related to an example mentioned above.

Below is a somewhat more detailed analysis of the risks under consideration, partly using the example of the printing product sketched above.

Table 5.5: Types of environmental risks as suggested by German Federal Banking Supervisory Authority (BAFIN).

	Physical risks	Transitional risks	Reputational risks
1	Operational risk	Default risk	Intern
2	Underwriting risk	Market(price) risk	Extern
3	Legal risk	Liquidity risk	
4		Strategic risk	
5		Stranded assets	

Market price risk/price development CO_2 levy

Following the decision to reform emissions trading in April 2018, the price of emission allowances tripled from an average of €5 per tonne of CO_2 in 2017 to €15 in 2018 [45]. In December 2019, the EU governments agreed to initially set the CO_2 price at €25 per tonne from January 2021. At the beginning of December 2020, the freely traded certificate price climbed to over €31, reaching a new all-time high. For 2026, a price corridor of at least 55 and a maximum of €65 is to apply a level already achieved in the free market duing summer 2021. The corresponding amendment to the law came into force after approval by the German Bundestag and will apply from January 2021. At the same time, the European Union (EU) announced that it would increase its climate target for 2030 from 40% to 55% compared to 1990. GHG emissions are thus to be reduced much more drastically in the coming years than previously planned. Higher prices are therefore quite possible – because the prices for emitting a tonne of CO_2 equivalent vary widely around the world. In 2020, they were as high as €140 per tonne of CO_2, in Sweden €140.

A report published in 2019 by the Berlin-based climate research institute MCC and the Potsdam Institute for Climate Impact Research (PIK) recommends that fixed CO_2 prices start at €50 per tonne of CO_2 in 2020 and then gradually increase to €130 in 2030. The Federal Environment Agency has proposed in an expert report that the price per tonne of CO_2 emissions be set at €180 in 2020. In order to limit global warming to 1.5 °C, the CO_2 price would have to be significantly higher, according to PIK estimates.

Development of CO_2 emissions

According to the Paris Climate Agreement (the UN Gap Report), global emissions must fall by between 3% (8%) annually by 2030 to limit climate change to well below 2 and 1.5 °C, respectively. According to the German Federal Ministry of Economics, CO_2 emissions fell to 35.7% until 2019 compared to 1990 – the target, as mentioned, is 55% by 2030. Total CO_2 emissions – from fossil CO_2 and from land use – were around 39 billion tonnes of CO_2 in 2020, still at the same level as around 2012, despite the decline [10]. In December 2020, emissions from road and air transport were between 10% and 40% below 2019 levels due to Corona-related constraints.

Assessment – monetary estimate – options for action

According to the above benchmarks and assuming that all countries worldwide work on their targets immediately, Germany needs to reduce GHG emissions by 2–4% per year (compared to the previous year) to reach target emissions in 2030. The most important instrument to achieve this is the commitment of companies to pay levers. Two approaches would be plausible for this – between which there could also be mixed forms:

1. Levy on total emissions with reference to a starting year: recording of total emissions and calculation of emission reductions.
2. Levy if the targeted reduction is not achieved, corresponding to the shortfall.
3. Probably the most likely variant: mixed form, with a basic amount paid on total emissions and a penalty levy if emission targets are not met.

Each of these approaches assumes that emissions are recorded continuously and reproducibly. Based on the above and a production site within the EU, a complex estimation for an installation results, depending on the assumptions:

1. It can be assumed that the authorities, at least in the case of large companies, will, due to the legal situation
   - evidence of continuous recording of GHG emissions,
   - evidence of the strategies pursued and their implementation or emission reductions.
2. The emission reduction target should be 4% or more reduction per year from today compared to 2019.
3. Legislators have no choice but to impose penalties to enforce the targets. As compliance checks for industries not subject to the ETS (including printers) will not be carried out until 2027, the corresponding penalties are not expected until 2027 or after, according to the current state of the law.

   Manufacturers who are already part of the ETS as suppliers will have to act earlier. Since for such manufacturers the emissions generated during production are attributed to the product, it is necessary to ask the manufacturers about corresponding strategies and the expected cost developments and to use them as part of one's own strategy or, if necessary, to change suppliers.
4. It can be assumed that the costs for corresponding penalties will be based minimally on the costs for CO_2 emissions – i.e. today €30, in 2025 at least €55–65/tonnes of CO_2 equivalent, but rather higher – currently a maximum of €180/tonnes of CO_2 equivalent. The extent to which these prices will remain unchanged cannot be estimated today – these are political targets. Quick or unexpected targets usually lead to equally quick price increases.

Example:
A printing company is currently involved in two stages of the value chain: a newspaper printed for its associated publisher and newspapers for other publishers. For the production of these printed products, 45,000 tonnes of CO_2 p.a. of emissions can be calculated.

The CO_2 footprint of the

- newspaper published by the parent company – the publisher – corresponds to about 40% of the print volume and
- other printed products correspond to the remaining share – about 60%.

The plausible idea from today's perspective for a legally enforceable development of emissions (and costs) could thus be handled as follows:

Table 5.6: Development of emissions and associated costs.

Year	2021	2025
Total emissions (CO_2 eq. t/a) including all print products Target: 3% reduction p.a.	45,000	39,800
Total scope 1 partly – addressable by the print shop	360	199
Total scope 2 partly –addressable by the print shop	1,260	1,115
Total scope 3 inbound – only addressable by partners in supply chain	36,450	32,240
Total scope 3 outbound – mostly in responsibility of partners (means of transport)) and depending on the location of own customers (assuming 100% recycling of paper)	7,200	63,670
Minimal price per t CO_2 emission in €	*~30*	*~55(65)*
Worst-case cost per t CO_2 emission in € Total cost for emission in Mio € Min Max	€180/t 1.35 8.1	€180/t 2.2 (2.6) 7.2

Note that it makes sense to allocate the costs to the polluters – these are mostly the internal processes and energy suppliers for scope 1 and scope 2 emissions and the companies involved in the upstream and downstream value chain for scope 3 emissions. This effect will also increase prices for customers. Ultimately, it is in everyone's interest to lower prices so as not to run the financial risk of discontinuing products or significantly reducing the profits made.

It is obvious that in this example scope 3 inbound (paper and, to a lesser extent, paper transport) is crucial. Any way to reduce this figure will lead to the greatest possible reduction in monetary risks.

The calculations were made under the assumption that 100% of the printed newspaper is recycled. Should recycling not take place to this extent, the CO_2 price for the non-recycled part must be added to the above calculation.

Measures needed to address this risk are as follows:

1) Strategies are needed to reduce emissions by the same 3%. The 3% reduction is to be understood as an example.
2) For all types of resulting measures, the cooperation of procurement and partner management is essential, especially for the reduction of scope 2 and scope 3 emissions, as these emissions cannot be realised alone. The comparison of the respective individual items determines the targets to be achieved.
3) It can be assumed that the partners for whom production currently takes place want to see emission reduction strategies on the part of the printing company.

The preparation of risk analyses such as this one with the corresponding figures tailored to the partner will therefore be necessary.

4) The associated costs are accommodated in the purchase prices and result in a lower profit or a higher price per product. The current legal position needs to be reviewed to ensure that the point at which emissions are tested (and direct costs incurred) is well understood.
5) Strategic risk

A printing company's carbon footprint is dominated by energy (gas and electricity) and raw materials (paper and consumables). The paper industry is known for its high emissions and is therefore – accordingly – part of the ETS. It is not unlikely that the changing environmental requirements will have an impact on both the price (increase) and the availability (reduction) of paper as a raw material. Already today paper for newsprint can be seen as a less lucrative production of raw material than (corrugated) board.

Credit risk – default risk

A printing company works in association with a paper supplier/supplies itself with paper. The paper supply chain has high emissions due to the technology used. These emissions are priced. The price development has been estimated above. Customers have to bear the corresponding costs for the products printed for them.

The risk is that the costs develop in such a way that the continuation of the printed product is no longer lucrative, which in turn leads to the cancellation of the print job. The same risk naturally applies to the entire value chain.

Valuation – quantitative assessment for customers

It is in the logic of the above approach that there are also increased costs for a printer's customers. These are most easily calculated per tonne of paper. It can be assumed that suppliers, in order to maintain their margin, will price in the levies (in particular the water, gas and electricity suppliers as well as the paper supplier, as they should all already be obligated to the ETS today) and pass them on as increased prices.

From a risk perspective, it should be considered that given the additional costs, some of the customers might not be willing to pay the resulting price. Consequently, they might decide to abandon and choose another partner who has higher margins or can take on a higher share of the costs incurred or can use a cheaper supply chain.

Reputational risk

As mentioned earlier, reputational risk can arise from both sudden events and long-term developments, and can have an impact on both the internal relationship with employees and the external relationship with partners and customers.

Table 5.7: Development of emissions and associated costs to be charged to the customers of the printing company under consideration, based on the data presented in Table 5.4.

Year	2021	2025
Total emissions (CO_2 eq. t/a) including products printed for partners (e.g. 60%) Target: 3% reduction p.a.	27,000	23,880
Minimal price per tonne of CO_2 emission in €	~30	~55(65)
Worst-case cost per tonne of CO_2 emission in €	180 €/t	180 €/t
Total cost for emission in Mio €		
Min	0.8	1.3 (1.6)
Max	4.9	4.3

Energy-related needs could be met in a number of ways – through the normal electricity mix or by a company zeroing out its own emissions through certificates. This means that the customer initially saves on emission costs, but in terms of reporting it

- depends both on the extent to which the supplier's emissions have been correctly reported as well as
- whether the measure reported as offsetting actually offsets the announced emission in the long term, and
- whether it would not have taken place anyway without the client's investment.

Strategic options for action

The quantified risks must generally be reassessed annually for reporting companies, whereby it is assumed that the implementation of reduction strategies is supported by consistent action. In the EU, audits are expected from 2026 to 2027, and to cover the foreseeable costs resulting from non-compliance with the requirements or mismanagement of the risks, damage provisions could be used as a tool for action.

Some of the risks discussed here are medium or long term in nature. Risks can be addressed rather early and proactively or a "wait and see" strategy can be pursued. In any case, hedging against the risks that can already be quantified today should be examined. Planning horizons should be broadened with a view to informed decision-making, taking sustainability risks and factors into account. Systematic processes for identifying, measuring, managing and reporting sustainability risks were implicitly proposed earlier. It is therefore recommended to

- critically analyse all risks,
- to exclude certain sustainability risks altogether, if necessary, and
- design the necessary measures.

In this context, it is also advisable to clearly communicate the handling of identified sustainability risks to one's own management, employees, customers and investors. In particular, any criteria for the exclusion or targeted management of certain risk positions should be presented externally in order to make one's own actions transparent to stakeholders and to avoid unsettling customers.

Conclusions

If the aim of an assessment is to provide practical access to carbon footprints and risks for companies, the scope 1–3 emissions as defined by the GHP are suitable for grouping and accessing the data and the resulting options for action. For access to sustainability risks, it is proposed to group the risks mentioned by BAFIN in a slightly modified way and to provide them with probabilities or scenarios in order to make their local relevance accessible and partly quantify them.

The examples given show that for a high-volume product that is recycled after use, the most important contributions to the carbon footprint are the raw material, the transport of the raw material (in this example, paper), product and the energy supply. As some material and energy suppliers – at least in the EU – are subject to the emissions trading system, it is expected that the price development of emission certificates will have a direct impact on the purchase prices of paper and energy. Strategies are needed to reduce emissions by 23% p.a. plus, which cannot be achieved by changing internal processes. Therefore, the cooperation of all partners in the supply chain is essential.

The most important risks to address are the development of raw material and energy prices and the strategic risks of the supply chain becoming unstable.

Table 5.8: Organisational and manufacturing contribution to the environmental footprint – European Directives.

Type of organisation involved	Characteristics	EU document for general guidance
Production of physical product	A physical product leads to some material with an LCA as method of choice to determine the environmental footprint	General Guide for Life Cycle Assessment – Handbook
Pure organisation	A pure organisation implies the focus on a carbon footprint	Organisational Footprint (OEF) Guide [46]
Mixed	Both models apply and results need to be merged – but depending on different aspects of the organisation involved	

Table 5.9: Generic representation of a simplified LCA – environmental footprint as energy and material calculation for the life phases of a product.

Phases in lifetime: resources and emissions	Energy	Materials ending up in products	Materials wasted
Raw material and semi-products			
Distribution			
Production of machine tooling			
Distribution			
Production of downstream products			
Distribution			
Scrapping and/or recycling/reuse			

5.3 Example: print on different substrates

The following is an example of estimating the environmental footprint resulting from an LCA for a material product [4, 21] – the production of printed products when printed on different media [53]. First, the environmental footprint is developed for various products that initially appear similar. As will be shown, despite the similarity of the products produced, the overall picture will only lead to different assessments of printing on different media when the entire value chain is considered. For the case considered here, the situation is as often in industry [5]: one product under consideration serves as a means of production for other products. Hence, not only the product itself (here: a small printing system) has to be considered in the balance sheets to be calculated but also the products upstream and downstream in the value chain (for downstream: prints in this example).

As stated above, the environmental footprint can be presented as an energy and material calculation:

At the beginning of the assessment, it is required to decide which phases in life cycle of the product to include in the LCA or how the results are to be scaled (per year machine life or per unit produced) – the decision on the boundary conditions applying.

Paper – one of the industrial materials considered – can be used as an example to illustrate the link between different phases of different life cycles and, in this case also to a circular value chain (Figure 5.1) [6].

The other main component of the process is the printing press; in its life cycle it is – based on incoming materials plus energy – manufactured, delivered, used and kept running through maintenance, dismantled in whole or in part, refurbished

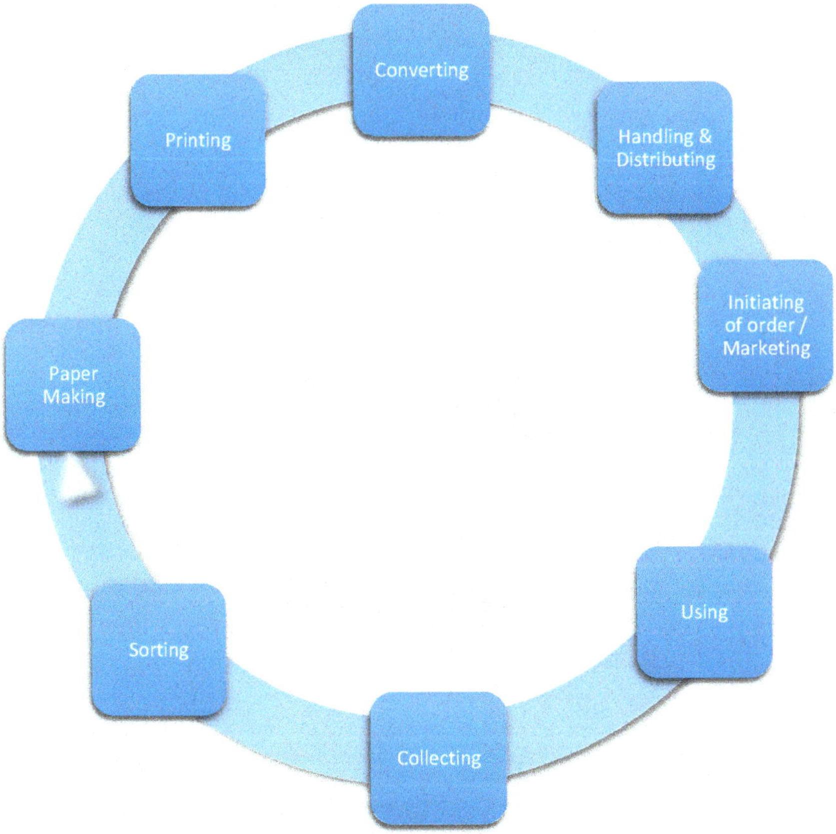

Figure 5.1: Circular value chain of paper in the context of printing [6]. The aim of the efforts involved in a life cycle analysis is to reduce the energy consumption and to keep the paper fibres in the value chain [7].

and delivered again to a (new) customer. A circular value chain is depicted in Figure 5.2.

The goal of LCA is, aside of reducing costs, to minimise consumption of energy and materials. That can be done by using less, different materials or extending lifetimes. This in turn may require trade-offs in other value chains associated with the consumables or supplies needed to run the press.

5.3.1 Boundary conditions

For a complete view of both the machine and the products generated with the machines, it is necessary to link the individual value chains into a more complex network. This case is illustrated in Figure 5.3.

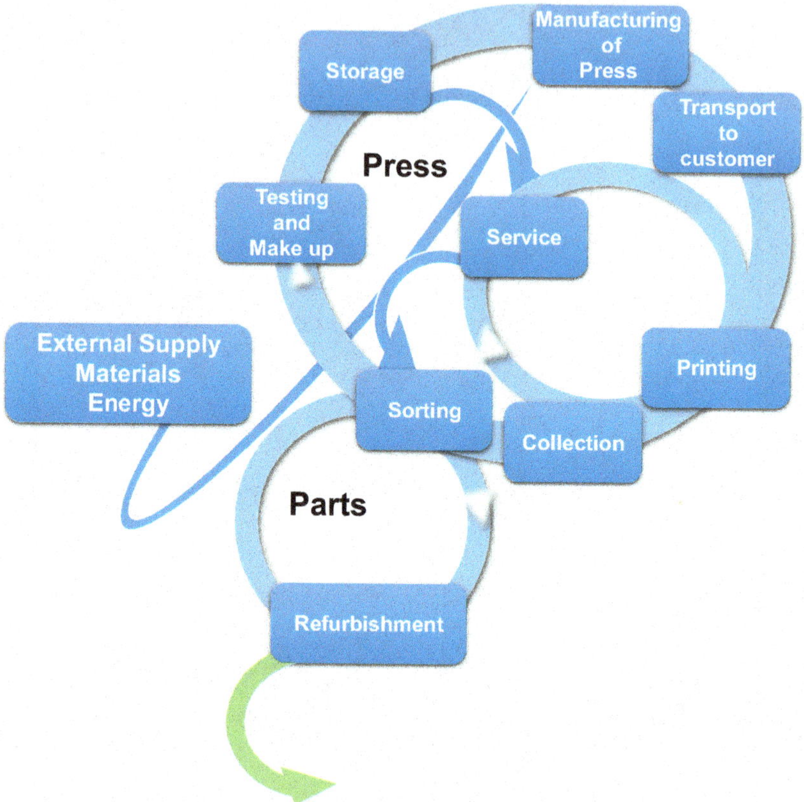

Figure 5.2: Circular value chain of a printing press.

As discussed above, the general phases of life to be discussed in principle are:

Raw material and semi-products	Distribution	Production of machine tooling	Distribution	Production of downstream products	Distribution	Scrapping and/or recycling/reuse

Even a figure like the one above seems too complex to discuss. It is helpful to define the phases in the life of the system/product and assigned them to what can be assessed about the life of the machine and the product generated (Figure 5.4).

As indicated above, essential parameters to be considered for the use cases of printing on different substrates vary with the materials used. These are, in particular, the energy and material resources consumed for production as well as the resulting waste and emissions during the various life phases. Since these values change with the lifetime of the press used, a list of influencing factors should also show how

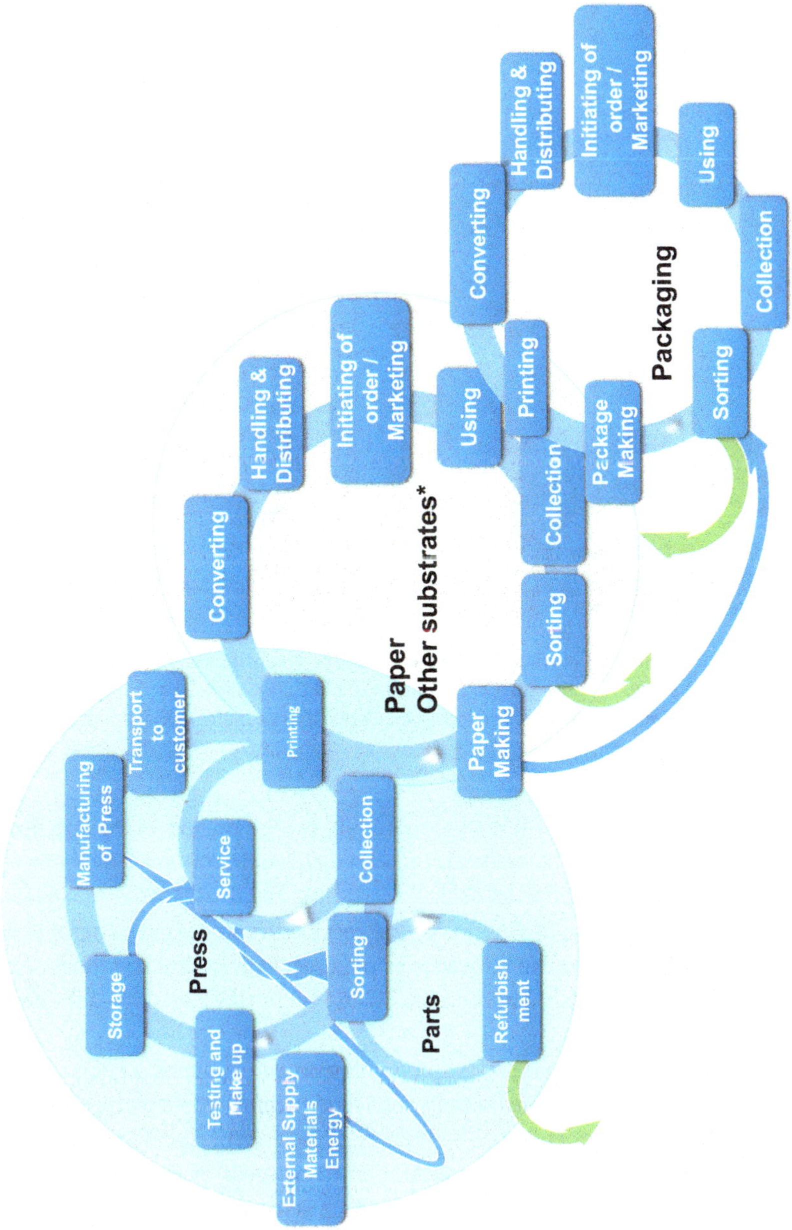

Figure 5.3: Circular value chain of the connected system of press and paper as substrate. It should be noted that a number of simplifications have already been made for this illustration and consumables are not shown in this figure.

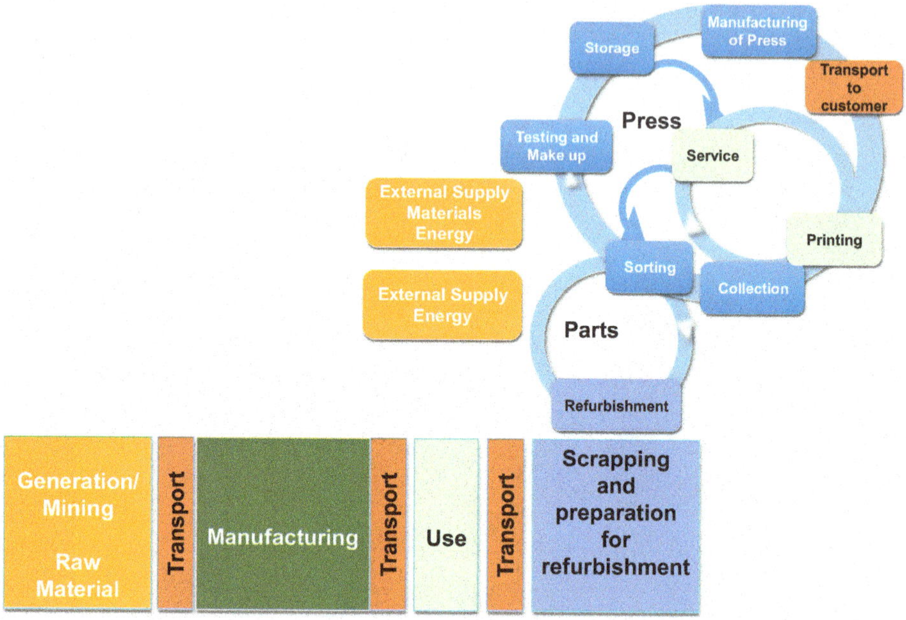

Figure 5.4: Linking the elements in the circular value chain of the composite system of press and paper as substrate to life stages. For the sake of simplicity, the footprint of the raw materials is only partially taken into account for this discussion. The colours chosen indicate the allocation of the individual phases.

presses are operated and to what extent the materials can be reused. Naturally, extending the service life has a positive effect on material consumption over the lifetime of a press. It therefore makes sense to design accordingly and to list the materials during the manufacturing phase or when they are recycled/reprocessed or scrapped as in the following list. Whether the substrate is paper, plastic or another does not lead to structural changes and is therefore not considered in the first step of the following discussion (Table 5.10).

The energy consumption for manufacturing the system was reported to be zero by the producer as renewable resources have been used to generate the energy [9, 10] – its manufacturer purchases and uses emission-free energy in the form of electricity and cooling/heating water (geothermal), as is now common in many areas of mechanical engineering. Direct emissions beyond the cooling water (which is recycled) are not known [9]. Moreover, simple rules can be applied for distribution. From a carbon footprint perspective, the transport phase should consider both the packaging and the system itself:

- Wood is needed for packaging, which is reused (see, e.g. [11]).
- Depending on the distance travelled, different transport methods can be studied and analysed. For this example, an average distance between user and production site

Table 5.10: Material consumption for recyclable materials (Al, Cu, Fe, certain plastics) and materials that can be reused to a large extent (according to 8).

Material type	Weight in newly built system	Fraction of total weight	Virgin material required for remanufacturing for second life	Reuse rate	Material consumption p.a. if lifetime is 7 years	Material consumption p.a. if lifetime is 11 years
Units	kg	%	kg	%	kg	kg
Printed circuit boards and batteries	2.5	1.36	0	0	0.36	0.23
Wire harnesses	1.4	0.76	0	0	0.2	0.13
Motors*	2.5	1.35	0.01	99.6		
Copper*	0.1	0.05	0.04	96		
Plastics*	34	18.48	3	90	0.42	0.27
Ferro*	111	60.32	17	85	2.42	1.54
Aluminium	10	5.43	4	58	0.57	0.36

For a new printer, for example, 17 kg more iron is needed than can be recycled at the end of its life. This means that the remaining share of iron must be divided by 7 or 11 to obtain 2.42 or 1.54 kg as annual consumption.

(1,000 km) and transport by truck was assumed and a web-based calculator was used [12]. This of course changes if transport by plane or ship is to be considered.

It should be noted that the issue of organisation of logistics and responsibilities is often not fully considered when discussing the reduction of environmental impacts of industrial goods (see, e.g. [13]).

This leads to the conclusion that the main factors that vary when printing on different types of substrates are (Table 5.11):

1. type of electrical energy (use of energy from renewable sources or not),
2. printing recyclable goods and using recycled paper (or other recycled/recyclable materials) and
3. extending the life of the printed product for a reduced footprint.

In the following sections, these aspects are considered for the cases of printing on paper, self-adhesive labels and directly on the object.

Table 5.11: Generic result of a life cycle assessment – environmental footprint as energy and material calculation for the lifetime of a product.

Phases in lifetime: resources and emissions	*Generation of raw materials (tonnes of CO_2 generated)*	Distribution	Production of machine tooling	*Distribution (tonnes of CO_2 generated)*	Production of downstream products	*Distribution (tonnes of CO_2 generated)*	Scrapping and/or recycling/reuse
Materials (kg)	*Disregarded/ unavailable*	*Disregarded/ unavailable*		0.2	Consumables Expendables Parts	0.2	
Energy	3.9 when not using reused parts 2.4 when using reused parts		0		Consumables Expendables Parts		
Other resources and waste	*Disregarded*		0				0
Emissions	*Disregarded*		–				–

5.3.2 Printing

Printing on paper

Lifetime of the printing press – usage life

For the type of press considered, it is primarily the absolute printing time that determine the consumption of substrates. The assumption of a second service life for certain parts (with simultaneous extension of the first service life of the entire system from, e.g. 7–11 years for both service lives) obviously also leads to a more positive balance. Extending the life of the system is made possible through design, planning and testing, as well as through increased use of preventive maintenance and upgrades throughout the life of the system. Parts are replaced during preventive maintenance and servicing and collected or, where possible, made available for refurbishment. If local life extension is intended, refurbishment can take place on site if the system is designed for it.

Usage scenario (printing operation) – energy

In order to assume a reasonable case for the consumption of the system under consideration, a usage scenario must be assumed. Such a scenario includes operating hours including switch-on time, warm-up time, service time, stand-by time and actual printing time. The power consumption for the different power settings is also needed. For certain systems, power consumption may also depend on what is being printed (e.g. depending on the chemistry of the ink used – if high viscous ink is used and coverage is high, more power is needed).

Application scenario – substrate

The type of substrate considered has an impact on the power consumption of the system. Also, the power consumption depends on the type of material to be printed but not on the origin of the energy. Moreover, while the absolute power consumption is a question of the system, the origin of the energy is not. Here, however, electricity consumers in most industrialised countries have a choice: while the proportion of climate-neutral electrical energy in the manufacturing phase can be assumed to be constant at 100%, the corresponding proportion at the customer's cannot be determined. This affects the assumed carbon footprint without the supplier being able to influence it.

As a consequence, consumption-related data for electrical energy can only be estimated, but not calculated as an absolute figure. For illustration purposes, a case is shown below.

Energy consumption can be described with a user profile and the duration of use. The energy consumption is the sum of usage during printing (~75% of the time), make-ready (~15%), energy saving (~10%) and sleep time when powered on. The carbon intensity value (kg CO_2/kWh) for electricity was provided by the IEA [14]. With a carbon intensity of 0.498 kg CO_2/kWh, the carbon emission thus corresponds to the one shown in Table 5.12.

Table 5.12: Energy consumption [9, 10].

Energy type	Carbon footprint (kg)
Electrical	763

For the analysis of the material flow during the use phase, the use characteristics are crucial and the same approach as described for the use of electricity applies. Furthermore, the technology used should be such that the printed substrate can be recycled.

The system must be designed in such a way that the print can be recycled without any problems (usually INGEDEs method 11 is recognised as the standard measure to describe deinkability as precondition to paper recyclability). In addition, the printing system must be capable of handling recycled substrate. The latter requirement is not self-evident and arises from the tendency of the short fibres in recycled paper to generate dust, which can cause nozzle clogging in inkjet printing.

The degree to which paper is made from virgin fibres or has been recycled and the prints can or cannot be subjected to recycling has a strong impact on life cycle energy consumption. Literature [16] shows that using 100% recycled paper instead of paper with 100% virgin fibres leads to a reduction in the carbon footprint of about 70%. The printed product can either be recycled or not. If it is recycled, fully collected and recycled, the lifetime energy consumption is also significantly reduced – but how much this effect matters depends on the type of printed product, the collection rate of the printed products and other factors. The calculation discussed here assumes that no recycled fibres are used and that the paper is not destined for recycling.

As with energy consumption, data related to paper or energy via paper consumption can only be estimated but not calculated as an absolute number. To illustrate this, a case is shown below.

We can look at best- and worst-case scenarios (from an eco-perspective) depending on whether the recycled paper can be used or not. The life of the press considered here be between 7 and 11 years. If the system can print on recycled paper, different carbon footprints offer the possibility to reduce the total carbon footprint that the system generates during its use phase. For the case considered, the savings potential is significant, as given in Table 5.13.

Accordingly, the Table 5.14 applies to paper.

The energy used for recycling is included in the energy footprint of the recycled paper. Consumables and parts were not taken into account as these parameters have no influence on the comparison of the substrates.

Table 5.13: Carbon footprint usage phase (electrical energy plus paper consumption) assuming (a) user profile as above and (b) 100% use of non-recycled or recycled paper.

Material type	Carbon footprint p.a. (kg eq.) if lifetime is 7 years	Carbon footprint p.a. (kg eq.) if lifetime is 11 years
Units	kg	kg
Paper	3,190	2,980
Recycled paper	1,970	1,840

Note that the absolute figures may also depend on the type of substrate used [9].

With this in mind, the main factors influencing the environmental footprint of printing on paper are given in Table 5.15.

Printing on plastic

In this context, a comparison of the data obtained for large format printers, as was done here with printing presses, leads to comparable results for both types of equipment [16]. The cited study also states that self-adhesive paper and polymers are comparable. Within the error window, the results reported here and by FINAT [16] are comparable (Table 5.16).

At first glance, the results for plain paper and self-adhesive labels (paper and polymers) are comparable. However, there are differences, especially since the recycling chain for polymer-based self adhesive labels hardly exists, "as recycled materials are rarely used as input for the production of label material" [16]. In other words, a comparable recycling chain as shown in the above figure "circular value chain of paper in the context of printing" does not exist for labels. In addition, recycling is established for paper, at least in Europe, but much less so for polymer substrates, as FINAT [16] also points out. FINAT suggests a sensitivity analysis for alternative waste treatment targets "when the actual waste treatment is unknown". For this reason, FINAT suggests as given in Table 5.17.

The absolute energy consumption for the production of polymers is between 15 and 25 kg CO_2/kg [14], while for paper it is about 10% of this [18]. While the amount of general plastics recycled in Europe is relatively low (~8.4% in 2017), recycling of some specific types is more significant.

The recycling rate of PET bottles and jars was 29.1% in 2017, while the rate for HDPE natural bottles was 31.2% [19]. For labels, the recycling chain is not as well developed as shown in Table 5.17. There are indications that the "pre-print waste" generated by the manufacturer of the plastics used for labels leads to an even worse balance [19]. The result for the environmental footprint for labels is given in Table 5.18.

Table 5.14: Generic LCA result – environmental footprint as energy and material calculation for the lifetime of a product for the case of printing on paper.

Phases in lifetime: resources and emissions	***Generation of raw materials (tonnes eq. of CO_2 generated)***	**Distribution**	**Production of machine tooling**	***Distribution (tonnes eq. of CO_2 generated)***	**Production of downstream products**	***Distribution (tonnes eq. of CO_2 generated)***	**Scrapping and/or recycling/reuse**
Materials (kg)	*Disregarded/ unavailable*	*Disregarded/ unavailable*		0.2	**Consumables** depending on the thickness of substrate Ink: 1–10 g/m^2 **Expendables:** neglected **Parts:** neglected	0.2	
Energy	3.9 when not using reused parts 2.4 when using reused parts		0		**Consumables** Paper 9.8 when not using recycled paper **2.9** when using recycled paper **Ink:** neglected **Expendables:** neglected **Parts:** neglected		
Other resources and waste	*Disregarded*		0		Wasted paper		0
Emissions	*Disregarded*		–		Depending on the type of ink – mostly steam		–

Note that ink has been neglected because for the system considered the substrate is recyclable and therefore the absolute waste for ink should be about 1–5% (of the weight of the paper to be recycled) for water-based inks.

Table 5.15: Main factors influencing the environmental footprint.

	Influencing factor
Energy	Consumption during production
	Consumption during use
	Use of renewable energy
Material *and* energy	Degree of utilisation of recycled parts
	Degree of utilisation of recycled paper and produce recyclable prints
	Ability to extend lifetime of all machinery involved
Material *strategy*	Choice of material (weight, generation of materials, value for and approach to refurbishment/recycling)
	Collection/disassembly strategy

Table 5.16: Comparison of the CO_2 footprint assessed for inkjet with the data reported by FINAT (for production with conventional presses). Within the limits of uncertainty, the results are comparable.

	FINAT focus: print converter and substrates	**This example:7 years lifetime (11 years)**	**This example: 7 years lifetime (11 years)**
	(%)	(% CO_2 footprint, rounded), excluding manufacturing	(% CO_2 footprint, rounded), including manufacturing
Electricity and heat	31	45–75 [30–50]	38–60 [50–75]
Label stock materials (note: 10–50% of label materials are waste according to [16])	63	25–55 [50–70]	13–71 [20–90]
Manufacturing of machinery	1	–	11–28 [10–20]
Packaging/transport	1	<1	<1
Printing inks	3	Neglected	Neglected
Waste	1	Neglected	Neglected
Water	0	Neglected	Neglected

Table 5.17: Sensitivity analysis of waste streams as proposed by FINAT [16].

Origin of waste	Waste streams to be considered	Default values waste	Default values treatment
Label stock manufacturer	Production waste	10%	Incineration without energy recovery
Printer converter	Production waste	10%	
	Matrix waste	40%	
Brand owner	Liner waste	100%	Municipal waste treatment (50% landfill and 50% incineration without energy recovery)
Post consumer waste	Label waste	100%	

Even if it were assumed that polymer labels were as easy to collect, sort and reuse as their paper counterparts (which is not the case), the carbon footprint for a polymer label would still be 5–10× higher than for a paper label. Plus, There are other factors to consider – most notably that paper and polymer labels are used in different applications with different functional requirements. For example, it may be necessary to laminate to protect the product, or that the polymer provides a barrier for food packaging. When calculating the material balance for paper and polymer labels, it is also important to consider that the percentage losses are different both in the production of the materials and in storage. The effect of the adhesive was not considered in this example.

Although a direct comparison is difficult, there is no evidence that printing self-adhesive polymer labels in any way results in a lower environmental footprint than printing paper. However, it must be stressed that the decision to use paper or polymer labels is not made solely on environmental grounds. Paper labels are not suitable to meet the requirements of certain applications and therefore the market is forced to accept the use of polymer labels in certain circumstances, however, negative the environmental comparison may sound.

Directly on objects

In principle, direct printing on objects can replace applied labels. From the aspect discussed here, reducing the environmental footprint in the context of an energy and material bill by printing on different surfaces sounds tempting at first. But under the aspect of an LCA that includes the material to be printed, the concept is no longer so promising. The concept becomes more attractive when it is understood more broadly and the phases of the life including possible collection of the object to be printed is included in the LCA.

Table 5.18: Generic result of a life cycle assessment – environmental footprint as energy and material calculation for the lifetime of a product for the case of printing self-adhesive polymer labels.

Phases in lifetime: Resources and emissions	***Generation of raw materials (tonnes eq. of CO_2 generated)***	**Distribution**	**Production of machine tooling**	***Distribution (tonnes eq. of CO_2 generated)***	**Production of downstream products**	***Distribution (tonnes eq. of CO_2 generated)***	**Scrapping and/or recycling/ reuse**
Materials (kg)	*Disregarded/ unavailable*	*Disregarded/ unavailable*		0.2	**Consumables:** depending on the type of substrate **Ink:** 1–10 g/m^2 **Expendables:** neglected **Parts:** neglected	0.2	
Energy 1,000 kg CO_2	3.9 when not using reused parts 2.4 when using reused parts		0		**Consumables:** highly dependent on the type of substrate – but no recycling possible. If equal base weight is assumed a factor 10 in carbon footprint. **Ink:** neglected **Expendables:** neglected **Parts:** neglected		
Other resources and waste	*According to FINAT [13] 2–3 times the waste generated by the converter*		0		Wasted labels In converting process: 50% (of weight including liner and matrix-related *waste)* [13]		0
Emissions	Neglected		–		*Depending on the ink used – solvents, monomers*		–

Direct labelling of objects seems promising, as no label material needs to be applied to the surface when printing. Assuming an existing approach to collecting, sorting, deriving a high and reproducible concentration of the substrate to be recycled and using it as a raw material, direct printing offers the possibility of saving the material for labels. As mentioned earlier, relatively few polymer containers are collected for recycling. In addition, direct-printed containers need to be cleaned and the print removed from the container, just as labels need to be removed.

Removing direct printing from containers is difficult due to often environmentally problematic chemicals used to remove the ink. In the case of printing on glass, issues would apply or the glass would have to be molten with the ink on its surface. The latter process would increase the ash content in the molten glass and thus reduce its strength – with the result that thicker glass may be required, leading to increased transport costs as a result.

Another aspect points to an additional problem: For certain applications, the label is considered a barrier, in addition to the material of the container. If the label is to be abolished, the barrier function must be taken over by the material of the container. This may be possible, but would change the packaging concept as such or require thicker containers with the same material.

However, there are also cases where printing directly on the object can provide a reasonable option from an energy and material point of view:

- If a reimbursement system is in place and no cleaning is required, then the problems outlined above would be avoided.
- If one assumes printing on the surface of industrial goods, then the approach seems promising.

All in all, if either no return is intended (such as with industrial signs) or the return policy allows access to well-sorted containers, then printing directly on the object is an initially promising scenario.

For this discussion, no clear rules could be found for the type of applications or the combination of materials used. Given the early stage of development, it is probably best to simply create a table of questions that should be answered before deciding whether direct-to-object printing is a promising idea, and further trials in advance of calculating a better LCA are shown in Figure 5.5.

5.3.3 Discussion

The methodology used to derive the environmental footprint as an energy and material calculation for the life cycle of a product is flexible enough to be applied to a situation where different life cycles are considered in the overall picture and for different use scenarios. For the cases evaluated here, the life cycles were divided into

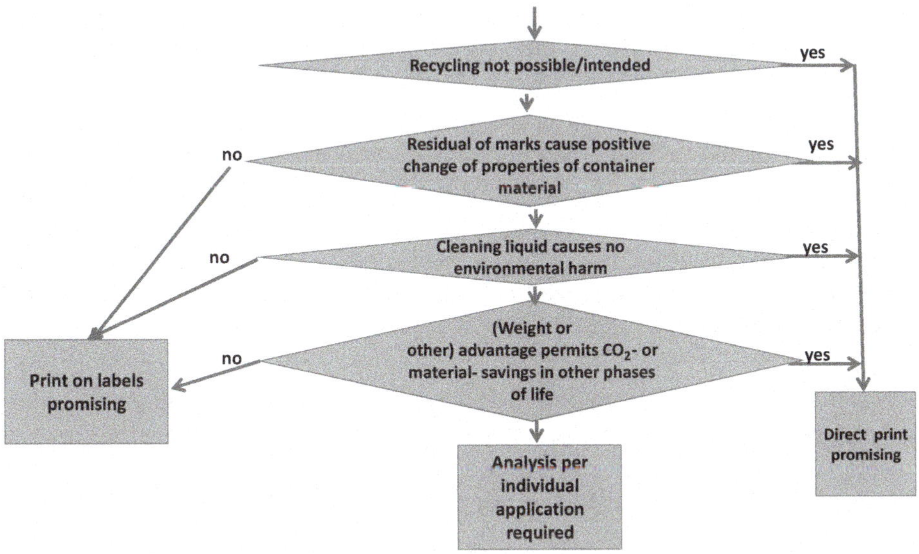

Figure 5.5: Guiding questions for the initial assessment of the likelihood that printing directly on the object is a promising idea to derive a better environmental footprint. Note that the arrow (no) of "cleaning fluid causes environmental degradation" strictly leads to a straight no if the cleaning fluid does not have an environmental clearance. The last two diamonds are not necessarily self-explanatory: The one labelled "Change of labels . . . " illustrates a case where printing directly on the object changes the packaging concept as such and thus implies a change of container. The new situation would require a new application and a comparison between the new and the old concept. The case marked "Rest of marking . . . " refers to a situation where the barrier properties of the container need to be re-evaluated due to the properties of the ink used in direct printing.

raw material extraction, manufacturing, transport, use, transport and repair/recycling. In addition, the effect of extending the service life was taken into account.

For the manufacturing/repair phase, access to relevant data for the analysis of industrial production proved to be difficult. For the type of plant considered here and the material footprint, the data can often be found in enterprise resource management or planning systems; for the carbon footprint, the approach based on scopes as defined in the GHP [17] was chosen. Furthermore, this example is based on data available from various sources. The combination of data gives an insight into the consumption patterns during the life cycle of a system in an abstract way – but may not apply to a concrete situation or product. The system serves as an example of a product type that is intended to be used to produce other goods. It was shown that the choice of material allows for high return rates, which limits the amount of material scrapped. For the use phase, energy consumption and the ability of the system to print on recyclable material were found to be the most critical elements.

From an eco-design perspective, it is clear that the following issues should be considered in the early stages of product design:

- the reduction of energy consumption during use,
- the ability of the system to print on recycled and recyclable material – e.g. paper (which is not given for certain labelling technologies),
- the markets in which it is acceptable to print on biodegradable substrates,
- the parts and materials strategy to enable reuse and remanufacturing as parts or as a raw material,
- a material strategy that avoids composites and variations of alloys to increase the value and recycling rate of product parts and
- the use of special design measures plus service to extend the life (of the system itself and the parts) and allow a second life (or more) after remanufacturing.

It was the scope of this example to highlight the effects of printing on paper, printing on labels and printing on objects. As long as it can be assumed that the systems enable printing on each type of substrate equally well, the machine part of the discussion can be neglected as it does not change the contribution of the substrate. Most importantly, the degree of recycling and sorting of printed products is key to successfully reducing the overall environmental footprint.

5.4 Example: backing oven

LCAs are often carried out in the first step in order to obtain an estimate of the contributions of the essential parameters in addition to meaningful definitions of boundary conditions. The following is an example of an assessment in such a rough approach – which will certainly be up to 30% different in detail for a specific appliance of the construction class. In this case, a standard household oven is considered.

First, the essential phases of the product's life are defined. Here,

1. raw materials obtained,
2. semi-finished products and the finished product are produced,
3. the product (baking oven) is used,
4. the product is disposed of and
5. transport between the individual use phases.

Assumptions for the estimation:

Ad 1

There are literature data for the distribution of materials in an oven: The report [47] gives a metal share of 77% for cookers and ovens in detail in Table 5.19.

Since the following is only a very rough estimate, it is permissible to limit it to the contributions of steel and aluminium and the averaged plastics. For a cooker

Table 5.19: Average material balance of cookers and ovens in %.

Material	Type of material	Percentage	CO_2 eq. per kg
Metals	Stainless steel	19.5%	1 kg stainless steel equals 1.34–1.67 kg CO_2 [48]
	Steel	54.3%	
	Aluminium	14.4%	1 kg aluminium equals 8–10 kg CO_2 [49]
	Copper	Not reported	
	Copper + aluminium Electronics	Not reported	
	Glass	Not reported	
Plastics	Acrylonitrile	1.8–1.9%	
	Butadiene	0.01%	
	Styrene	0.51%	
	Polystyrene, polyamide, polycarbonate, polyethylene, polypropylene, polyurethane foam, polyvinyl chloride	0.19%	
	Other plastics	1.7%	
	Other	5.6%	

with an oven, a weight of about 40 kg can be assumed. This results in Table 5.20 for the approximate CO_2 equivalent.

Table 5.20: Average material balance of cookers and ovens in absolute contributions for an oven having a weight of 40 kg.

	Mass contribution to the cooker	CO_2 equivalent kg eq. CO_2 per kg material	Resulting equivalent of the raw material kg eq. CO_2
Iron	30 kg	1.7	51
Aluminium	6 kg	8–10	48–60
Non-ferrous materials	4 kg	50–70 [50]	200–280
Total			300–370

Ad 2

The production is difficult to estimate, especially because it depends on the type of energy supply (gas, electricity mix) of the contributing companies. If the other devices presented in this lecture are taken as a basis, a contribution of at least 10% to the total CO_2 footprint over all life phases can be expected.

Ad 3

During the use phase, the oven is only used to produce food and, if the appliance has such a mechanism, for pyrolysis. The existing literature data do not provide any information on pyrolysis. With that caveat, a literature analysis provides:

1. for ovens, the annual electricity cost for an efficiency class A appliance is on average about 165 kWh [51],
2. for the year 2019, 401 g of CO_2 per kWh [52],
3. a service life of about 15 years can be assumed.

With these framework conditions, a value of 992 kg eq. CO_2 follows for the emissions during the entire life phase. For comparison, the Ökoinstitut [28] provides emissions for baking and cooking of 224 kg eq. CO_2/year. This means that for a lifetime of 15 years, 3,360 kg eq. CO_2 follows. If the two values roughly match, this corresponds to almost 2,400 kg eq. CO_2 for cooking alone during the lifetime of the appliance.

Ad 4

For disposal, refurbishment cannot be assumed. Here, it is assumed that disposal "costs" a minimum of 20% of the material-related emissions produced during manufacture, which leads to a burden on the CO_2 balance of 60–70 kg eq. CO_2.

Ad 5

Transport costs are often in the lower single-digit percentage range – an estimate that is in the right order of magnitude for this type of product. Here, 5% of the total balance is applied.

This leads to the estimation for overall balance for the baking oven given in Table 5.21.

Obviously, this estimate contains errors – because a cooker and an oven are not the same thing, the extent to which an average emission of 990 kg eq. CO_2 is realistic in all regions of use and household types (usage profile, number and habits of household inhabitants) can very well be questioned. Nevertheless, the orders of magnitude and ratios are reasonable and point to conceivable priorities in eco-design:

- The energy mix used during the use phase (or the use of the energetically more efficient natural gas).
- Improved insulation of the oven.
- Reuse of the baking heat after the end of the baking process.

Table 5.21: Total CO_2 balance for the oven.

	Phases in lifetime	Estimated percentage equivalent of life stage	Resulting equivalent of the life phase in kg eq. CO_2
	Raw materials		300–370
	Production	10% minimal	160–170
	Use phase		990
	Waste		60–70
	Transport	5%	~ 80
Total (rounded)			~ 1,490–1,680

- The abandonment of pyrolysis or its replacement by more energy-saving cleaning methods.
- A flexible design of the baking chamber volume to avoid a large heated volume.
- A temperature control of the baking process designed for energy efficiency, e.g. by programming depending on the weight or fat or water content of the baked goods and the recipe used or by sensory detection of essential characteristics of the baked goods.
- Energy optimisation for special forms of use (drying, fermenting and keeping warm). Independently of this, other measures to save emissions are also urgently needed.
- Simplifying the dismantling of the cooker/bake oven.
- Use of alternative heating methods (e.g. mixed operation with timed use of microwaves or also (focused) IR radiation).
- Modified raw materials in order to simplify the reuse of the materials.

In principle, however, it should be noted that the above approach outlines the perspective of the cooking appliance, albeit inaccurately. From the perspective of an LCA of the food, it would also be necessary to ask what share the stove/baking oven has of the total energy demand of the food
- contribute to the total energy demand of the food,
- can contribute to the reduction of food waste,
- can have on the quality of the food produced.

Or to what extent ONE type of cooker/baking oven should be configured in the same way for very different types of use and regions, even if they are of comparable design.

What this example was intended to show was that even a relatively rudimentary analysis of the available data often provides initial indications for an analysis and promising approaches to CO_2 reduction – which should not discourage a detailed analysis, because the error in the above approach may well be in the double-digit percentage range, but the proportions should be correct. However, this approach neclects that an oven is operated in the context of other heat emitting and heat consuming devices. Re-using the heat coming from a refrigerator or using the heat coming from an oven in the kitchen wold be reasonable.

5.5 Example: digital newspaper

The following model was originally calculated in order to develop a model and to obtain indications for weighing up the extent to which a switch from an analogue to a digital magazine is preferable from a consumer's point of view in terms of the CO_2 emissions caused. Since no concrete newspaper was used as a comparison, literature data had to be assumed, as in the previous example. This is a weakness of the approach because, as far as several crucial parameters are concerned, only one literature source could be identified, which ultimately makes a plausibility check impossible.

As shown in the example above, the approach is stepwise: After the individual life phases of the digital product have been described and the framework conditions of the analysis have been defined, the emission equivalents are calculated according to the definition of scopes 1, 2 and 3 of the GHP based on literature data.

5.5.1 Phases in lifetime

In the following, we consider of four life cycle phases:

1. Programming of software
2. Production of data (= articles, videos)
3. Distribution and storage of data in server farms
4. Reading/viewing/consumption

of the analogue newspaper, which are first considered and then integrated into an overall picture. We do not have a specific newspaper in mind, but assumptions about circulation have to be made to ensure that the result has some realism. These figures have been chosen so that they could be realistic for a large and nationally run newspaper. We also assume that:

- The software used is standard – software used by many clients worldwide and therefore the programming is NOT done for one publisher only.
- The digital publication is created and edited by a team of 400 (part- and full-time) reporters.

- The server farm used is not for one digital product – but we do not know how many digital products are hosted there and how many people are needed or employed to run the environment.
- e-Paper and internet-based publications are the same – no matter where they are consumed.
- The number of users/readers or consumers plus the time they spent reading on the newspaper's website is a crucial component – but by no means the same for every newspaper – again assumptions were made.

Against this background, a mix of data was used to work out some effects – for a specific practical case, this should make the model meaningful, but not the individual parameters.

Editorial Team/Reporters

For the sake of plausibility, it can be assumed that equipping a publishing house with 400 reporters with standard software for daily use requires a common number of programmers that is significantly smaller than the number of reporters equipped. The newspaper's readers also use standard software. Software is a mass-market product. Thus, an assumption can be made regarding the number of people involved (programmers, service, etc.) and computers used for programming, testing and maintenance. Here, an equivalent of 10 people and 100 computers was needed for all the software to be found on the desktop of the reporters or readers in connection with this usage scenario. This should not exceed an equivalent of about 2.5% of the total consumption given in above (1,055,280 kg eq. CO_2) – i.e. significantly less than 26,132 kg eq. CO_2.

As indicated, approx. 400 reporters (including back office and support) are assumed. They work 8 h+ per day and use standard software. They travel little for reporting and largely use video conferencing with their contacts. We assume that these reporters have a laptop and a smartphone in permanent use – where permanent means 18 h a day. We further assume that the hardware used by the reporters is no different from that normally used by private users.

Heat and office

We assume a CO_2 consumption of 0.91 tonnes of CO_2 per employee per year. The number considers for heating and water [36].

Additional occupational activity

We also assume that a reporter produces a large number of articles, photos and videos that are considered comparable in terms of consumption when shipped abroad to "50 g eq. of CO_2 for one with a photo or a hefty attachment". However, these figures were generated well over 10 years ago and are likely to be larger today. We assume that on average 30 of these emails are sent/received per day – resulting in an outgoing transport of 1.5 kg eq. CO_2 per day. It is known that the newspaper has about 400 people who deliver the news (or the corresponding infrastructure). The data published in [30] was generated under the assumption that the laptop is used 4 h a day with an electrical power consumption of 32 W. The smartphone also hangs on the charger for 4 h a day and consumes 5 W during this time.

We also assume 2 h of video conferencing per day, which would produce 2–107 kg eq. CO_2 and kg CO_2 eq. [27].

5.5.2 Server farms

Server farms are used for receiving, transmitting, storing data and related traffic. It is assumed that a server farm hosts more than one digital product.

5.5.3 Read/consume

Access to the data of specific server farms is not possible. Therefore, estimates have to be used. For this purpose, the data in the publication [25] refer to a user operating a laptop and a smartphone. The Federal Association of Digital Publishers and Newspaper Publishers [29] speaks of 35–40 min of average usage/consumption time of digital newspaper per day. We further assume that the usage time per day for a laptop is 3–4 h (according to [30]) and a smartphone is probably not switched off at night.

The difficulty with the underlying assumption lies in the design of the website in question and especially the use of videos – several reports point out that streaming video alone consumes about 1/4 of the internet's electricity [31, 32]. It generates 300 million tonnes of carbon dioxide per year, which is about 1% of the global emissions, according to the French Shift Project. According to this source, this is due in particular to the fact that, in addition to the electricity consumed by the devices, the servers and networks that distribute the content also consume energy (quoted after [39]). Apparently, data-intensive images or advertisements also contribute to the GHG consumption of server farms (screening life cycle assessment of printed, Web-based and tablet-e-paper newspapers [44]). Consequently, a detailed analysis of the respective consumption data would be necessary to derive an accurate carbon footprint (Table 5.22).

Table 5.22: Total emissions arising in connection with the publication of a digital newspaper.

Coarse estimation					
Scope 1					
No.	**Category**	**CO_2 emissions in kg eq. CO_2 per 10 programmers/reporter/user**			
		Programming	**Reporters**	**Server farms**	**Reading**
	Direct emissions	Not available	Not applying	Not available	Not applying
Scope 2 (12.2% for analogue, based on some assumptions 67–71% for the digital publication)					
No.	**Category**	**CO_2 emissions in kg eq. CO_2 per 10 programmers/reporter/user**			
		Programming	**Reporters**	**Server farms [12]**	**Reading**
2	Indirect emissions	4,700 [1,2]	1,100 [2, 10]	36–42	155.8–167 [1, 12]
Scope 3 (In-/Outbound, 87% for analogue, 29–33% for digital)					
No.	**Category**	**CO_2 emissions in kg eq. CO_2 per 10 programmers/reporter/user**			
		Programming	**Reporters**	**Server farms [12]**	**Reading**
3.1	Computers/smart phones/router	3,500 kg/ year [3]	710 [11]		63.6–76.2 [12]
3.2	Means of production/capital goods are all included in line 3.1				

(continued)

Table 5.22 (continued)

Coarse estimation							
3.3	Fuel and energy-related emissions	9 [4]	11.7 [4]				
3.4	Upstream transport and distribution	66 kg [5]	6.6 [5]	No data available		6.66 [5]	
3.5	Consumables	1,000	200	No data available		Not applying	
	Waste	1,000 [6]	Not applying	No data available		Not applying	
3.6	Business trips	4,000 [7]	1,200 [7]	No data available		Not applying	
3.7	Commutes of workers	400 [8]	200 [8]	No data available		Not applying	
	Total scope 1-3	2,638.2	36–42	235.1–260.0		235.1–260.0	
	Total scope 1-3 per publisher (assuming ~ 300 000 readers and 1.6 mio digital daily readers)	13,860	1,055,280 400 reporters	57,600,000 35 min of reading	67,200,000 40 min of reading	376,160,000 35 min of reading	416,000,000 40 min of reading

5.5.4 Discussion

As already mentioned at the beginning, the above estimate is very high level – especially because only one source was available for the energy demand of server farms. An error of 30% in the final result is therefore very likely – nevertheless, the result should correctly reflect the magnitudes of the individual contributions. Relevant here is that

- the break-even goes in favour of the analogue newspaper from a reading time of a few minutes onwards,
- the programming and editorial phases are relatively low or even negligible in terms of their contribution to emissions,
- the energy consumption of servers and reading infrastructure represent the largest part of the emission-triggering factors – and can be influenced by the energy mix.

As far as break-even is concerned, KTH Stockholm published a study in 2011 [49]. It concluded that "with a reading time of 30 min per day the environmental impact of the web based newspaper was often in the same range as the printed newspaper environmental impact". In this example, however, server farms and routers were not taken into account the like the fact that per analogue newspaper about 2.4 readers can be assumed. With the corresponding conversion, this would lead to a break-even between analogue and digital newspapers of about 8 min – i.e. according to this much more precise calculation, the analogue newspaper would have a lower emission than the digital newspaper with a reading time of about 8 min. If the reading time were reduced to about 1.5 min per day in the coarse estimate presented here, the digital newspaper and the analogue version would be about equally attractive.

There are publications that point out that the design of the digital newspaper's website, the number of videos inserted, the design of the software used have a considerable influence on the break-even – this was not taken into account in this artificial example.

5.6 Summary

As shown, a step-by-step approach can be taken to determine an LCA. Here we used:

1. Model generation for the product/organisation under view (boundary conditions, phases in lifetime of product, literature research)
2. Assessment of data internally available and data from up- and downstream supply chain
3. Set up of balance sheet
4. Fill in data and, for what is not available internally, literature values or estimation

5. Analysis of data – impact analysis
6. Risk assessment
7. Identification and priority setting for actions
8. Planning of actions

B. Murray of the Product Sustainability Forum in the UK [24] has summarised the same procedure as given in Table 5.23.

Table 5.23: Steps for the eco-design of a material product according to the Product Sustainability Forum, UK (see also [24]).

Step 1 Screening of assessments (such as emissions according to the GHG Protocol or materials used in processes) and subsequently for all stages of the product life cycle: – Context of the product including sales volumes – Identification of potential environmental "hotspots" – Identification, selection and prioritisation of sustainable design principles and assessment criteria
Step 2 Generate and compare design alternatives for all phases of the product life cycle For example, linked to the – Development of a scoring system to evaluate hotspots and possible solutions – Develop scenarios for production, use and disposal and analyse the interactions between the phases – Design eco-design strategy wheel [22] – Rigorously monitor cost impacts for all scenarios to obtain life cycle costing
Step 3 Review Progress and impacts are reviewed Key factors to assess at this stage include impact on: – Product quality and performance and sales – Costs and savings – Impact on manufacturability and organisation – Impact on environmental savings – Impact on brand equity
Step 4 Develop introduction and communication plan The basic concepts of the added value and environmental benefits of the product or service are communicated to stakeholders or included in prescribed declarations [23]
Step 5 Establish an eco-design strategy in conjunction with brand planning – Eco-design principles must be anchored in the corporate strategy – Specific eco-design criteria are developed after hotspots are identified – Principles include existing targets plus commitments, audits, standards followed, certifications, e.g. related to consumption, footprint or raw material use or waste or packaging reduction

Comparing the accuracy of the different examples given leads to the insight that the more a footprint analysis is based on literature the less accurate it becomes – not only for a given practical case but also in general. This is not only due to the fact that for any given real-world production environment, literature data are less accurate compared to the actual invoices to be taken into account but also literature data are most often not as recent as the information on the local environment. On the other hand, most often certain data are not available – e.g. in the newspaper case the CO_2 footprint of the concrete ink used or the CO_2 footprint of the metals actually used in an oven.

In addition, comparing the different examples also gives reason to believe that for alike products it is highly probable that

- in the balance sheet the same factors can be neglected and
- the relative weight of contributing factors is comparable.

This rule of thumb is used by several internet-based calculators for footprints. These often base their calculations not on the full (i.e. all scopes fully considered) but on a reduced set of parameters out of the scopes to be analysed. While that may work occasionally, it may not be justified in others.

If the target is to analyse a given product/production environment/company and to derive strategies and actions to be taken to reduce footprints for a given product ("eco-re-design") or with eco-design for a new product, it is most often advised to start with a model for the phases in lifetime and proceed by assessing all data available, for each phase, with considering all scopes for each phase. In parallel, it is reasonable to also assess the risks going along with the footprints as they may also heavily impact the strategies derived and, consequently, the resulting actions and their sequence.

In the early stages of an eco-design, it is important for practical reasons to keep spirits high – so it is advisable to motivate a team. As success is the best motivator, it is helpful to be able to identify "quick wins". Without limiting or restricting the activities that lead to a more detailed analysis, it is useful to consider the following aspects to access "low hanging fruit" (Table 5.24).

Table 5.24: Common first steps in eco-design.

- Procurement electricity from zero- or low-emission sources
- Sharing (access to) resources with neighbours (also question of sharing access to heat and waste heat)
- Buying and using recyclable and recycled material
- Monitoring/describing materials entering/leaving the production site and splitting flows into product and non-product-related information
- Analysis of competitors and which markets they serve
- Identification of relevant environmental/sustainability claims on the part of which stakeholders
- Where possible, assess and implement the product's ability to use (and produce) recycled and recyclable materials

It is crucial to set the boundaries right at the beginning of the approach as to what is and is not included in the example – in particular whether and which supply chain is included and whether one's activities are upstream or downstream. As was visible in step 3 above, stakeholders – the parties within the company that hold various responsibilities of importance for the product – are asked to participate and contribute to all activities.

References

[1] Baumann, H., Tillman, A.M., 2004. The Hitch Hiker's Guide to LCA. An Orientation in Life Cycle Assessment Methodology and Application. Lund, Sweden: Studentlitteratur.

[2] European Commission, 2010. Making Sustainable Consumption and Production a Reality. Luxembourg: Publications Office of the European Union

[3] European Commission – Joint Research Centre – Institute for Environment and Sustainability, 2010. Carbon Footprint – What It Is and How to Measure It. Luxembourg: Publications Office of the European Union

[4] European Commission – Joint Research Centre – Institute for Environment and Sustainability, 2010. International Reference Life Cycle Data System (ILCD) Handbook – General Guide for Life Cycle Assessment – Detailed Guidance. Luxembourg: Publications Office of the European Union

[5] Goedkoop, M., De Schryver, A., Oele, M., 2008. Introduction to LCA with SimaPro 7. Netherlands: Pré Consultants

[6] ISO 14040:2006. Environmental management – Life cycle assessment – Principles and framework. International Organization for Standardization

[7] PAS 2050:2008. Specification for the assessment of the life cycle greenhouse gas emissions of goods and services. BSI

[8] SAIC, 2006. EPA 101: Life Cycle Assessment: Principles and Practice, Ohio: US Environmental Protection Agency.

[9] UNEP, 2004. Why Take a Life Cycle Approach? Paris: United Nations Environment Programme.

[10] UNEP, 2007. Life Cycle Management – A Business Guide to Sustainability. Paris: United Nations Environment Programme.

[11] Weidema, B., Thrane, M., Christensen, P., Schmidt, J., Løkke, S., 2008. Carbon Footprint – A Catalyst for Life Cycle Assessment? Journal of Industrial Ecology, 12 (1), pp. 3–6, February 2008

[12] Wiedmann, T., Minx, J., 2007. A Definition of 'Carbon Footprint'. Durham, UK: ISA UK Research & Consulting (ISA UK Research Report 07-01)

[13] Society of Environmental Toxicology and Chemistry (SETAC), 1990

[14] National Risiko Management Research Agency, Life Cycle Assessment Principles and Practice, https://nepis.epa.gov/Exe/ZyNET.exe/P1000L86.TXT?ZyActionD=ZyDocument&Client=EPA&Index=2006+Thru+2010&Docs=&Query=&Time=&EndTime=&SearchMethod=1&TocRestrict=n&Toc=&TocEntry=&QField=&QFieldYear=&QFieldMonth=&QFieldDay=&IntQFieldOp=0&ExtQFieldOp=0&XmlQuery=&File=D%3A%5Czyfiles%5CIndex%20Data%5C06thru10%5CTxt%5C00000002%5CP1000L86.txt&User=ANONYMOUS&Password=anonymous&SortMethod=h%7C-&MaximumDocuments=1&FuzzyDegree=0&ImageQuality=r75g8/r75g8/x150y150g16/i425&Display=hpfResearchandDevelopmentefSeekPage=x&Search

Back=ZyActionL&Back=ZyActionS&BackDesc=Results%20page&MaximumPages=1&ZyEntry=1&SeekPage=x&ZyPURL

[15] "IMPACT 2002+" LCIA methodology / Dr. Olivia Jolliet, Univ. of Michigan, http://www.sph.umich.edu/riskcenter/jolliet/impact2002+.htm, Retrieved April 10 2020

[16] Environmental Management – Life cycle assessment – Requirements and guidelines, www.iso.org/standard/38498.html

[17] European Commission – Joint Research Centre – Institute for Environment and Sustainability: International Reference Life Cycle Data System (ILCD) Handbook – General Guide for Life Cycle Assessment – Detailed Guidance. First edition March 2010. EUR 24708 EN. Luxembourg. Publications Office of the European Union; 2010, http://lct.jrc.ec.europa.eu/pdf-directory/ILCD-Handbook-General-guide-for-LCA-DETAIL-online-12March2010.pdf/view, Retrieved April 10 2020

[18] http://www.scienceinthebox.com/en_UK/sustainability/casestudies_en.html, Retrieved April 10 2020

[19] www.unilever.com/sustainability/environment/manufacturing/assessment/index.aspx, Retrieved April 10 2020

[20] www.vestas.com/en/about-vestas/sustainability/wind-turbines-and-the-environment/life-cycle-assessment-%28lca%29.aspx Retrieved April 10 2020

[21] www.bath.ac.uk/research/seminars/esrc/lca/BathJuly2008NU.pdf Retrieved April 10 2020

[22] https://en.wikipedia.org/wiki/BP

[23] https://ghgprotocol.org/sites/default/files/standards_supporting/FAQ.pdf)

[24] https://ghgprotocol.org/sites/default/files/standards_supporting/Kapitel1.pdfhttps:/ghgprotocol.org/sites/default/files/standards_supporting/Kapitel1.pdf

[25] https://www.oeko.de/oekodoc/1029/2010-081-de.pdf

[26] https://observablehq.com/@mrchrisadams/how-do-i-work-out-the-carbon-footprint-of-providing-space-to-w, Zugriff 9. Mai 2021

[27] https://www.bbc.com/future/article/20200305-why-your-internet-habits-are-not-as-clean-as-you-think-Zugriffam11. Mai 2021).

[28] https://blog.oeko.de/digitaler-co2-fussabdruck/-Zugriffam11.05.2021

[29] Zeitungsmarktforschungsgesellschaft (ZMG), zitiert von Andrea Gourd (gourd@bdzv.de), +49 69 97382246 Anruf am 26.04.2021

[30] https://www.quora.com-Zugriffam11. Mai 2021

[31] http://gauthierroussilhe.com/en/posts/digital-sustainability-a-french-update

[32] https://greenit.fr

[33] https://theshiftproject.org/wp-content/uploads/2019/03/Lean-ICT-Report_The-Shift-Project_2019.pdf

[34] https://at.twosides.info/wp-content/uploads/sites/7/2011/09/07-12-Life-cycle-of-print-and-e-media-KTH-Stockholm.pdf

[35] https://theshiftproject.org/wp-content/uploads/2019/03/Lean-ICT-Report_The-Shift-Project_2019.pdf

[36] https://observablehq.com/@mrchrisadams/how-do-i-work-out-the-carbon-footprint-of-providing-space-to-w Accessed May 9th 2021

[37] https://i.dell.com/sites/content/corporate/corp-comm/en/Documents/dell-laptop-carbon-footprint-whitepaper.pdf)

[38] https://de.wikipedia.org/wiki/Übertragungsverlust

[39] https://theshiftproject.org/wp-content/uploads/2019/03/Lean-ICT-Report_The-Shift-Project_2019.pdf

[40] https://cundall.com/Cundall/fckeditor/editor/images/UserFilesUpload/file/WCIYB/IP-6%20-%20CO2e%20emissions%20due%20to%20office%20waste.pdf

[41] https://www.greenit.fr/2008/10/17/50-des-emissionsde-co2-liees-aux-voyages-d-affaires/
[42] https://www.amoes.com/media/societe/impact-carbone/bilan-carbone-amoes-2017.pdf
[43] https://librairie.ademe.fr/cadic/3541/ademe-ges-tic-0212.pdf
[44] https://at.twosides.info/wp-content/uploads/sites/7/2011/09/07-12-Life-cycle-of-print-and-e-media-KTH-Stockholm.pdf, Seite 92
[45] https://www.bafin.de/SharedDocs/Downloads/DE/Merkblatt/dl_mb_Nachhaltigkeitsrisiken.html
[46] https://ec.europa.eu/environment/eussd/pdf/footprint/OEF%20Guide_final_July%202012_clean%20version.pdf
[47] http://www.materialflows.eu/assets/Material_Flows_of_the_HA_Industry_LR.pdf, Seite 33
[48] www.Edelstahlrohrshop.com
[49] www.wir-leben-nachhaltig.com
[50] https://www.sciencedirect.com/science/article/pii/S2352484720315985
[51] https://www.net4energy.com/de-de/smart-living/backofen-stromverbrauch
[52] https://www.umweltbundesamt.de/presse/pressemitteilungen/bilanz-2019-co2-emissionen-pro-kilowattstunde-strom
[53] Has, M., Sustainability and Eco Footprint – Concepts for the application of Circular Economy Criteria for printing systems and their use when printing on paper, labels and direct-to-object, Chapter 19.1 "Ecological Aspects", in: Zapka, W. (Ed.), "Inkjet Printing in Industry", Wiley-VCH, 2022.

Chapter 6
Eco-design

Principles of sustainability include not only environmental but also economic and social aspects. As mentioned above, over 80% of all product-related environmental impacts can be influenced during the product design phase [7]. According to [8], "product design" means "the set of processes that translate legal, technical, safety, functional, market or other requirements for a product into the technical specification for that product". Sustainable design (often referred to as eco-design, green design or environmental design) is an approach to the design of products. "Eco-design" refers to "the integration of environmental aspects into product design with the aim of improving the environmental performance of the product throughout its life cycle" [1, 8].

6.1 Guidelines

EU Directive 2009/125/EC establishes a framework for setting eco-design requirements for energy-related products. Member States are required to "lay down the rules on penalties applicable to infringements of the national provisions adopted pursuant to this Directive and to take all measures necessary to ensure that they are implemented". The penalties provided for must be effective, proportionate and dissuasive, taking into account the level of non-compliance and the number of units of non-compliant products placed on the Community market" [1]. In general, eco-design directives facilitate the circular economy by encouraging the re-use of materials already obtained rather than the use of new resources. However, there is no general advice on how to recover such materials in a way that does not compromise the properties of the recycled materials and how to manufacture them with efficient energy use and minimal waste generation. The basic idea is to reuse already recovered/prepared materials for the production of new products. In addition to the actor-specific guidelines, general eco-design recommendations apply to each actor in the life cycle of the product. The general mind-set should be that all actors consider the value of the product by striving to provide the most recyclable solution and minimal degradation. The recommendation is to minimise system-level impacts and negative externalities, regardless of the source, and commit to continuous improvements in this area.

In particular, resource efficiency is a key principle that should be implemented upstream of sustainable product design development; resource efficiency aims to reduce impacts at all life stages by minimising consumption and enabling resource recovery and reuse. See Table 6.1.

https://doi.org/10.1515/9783110767308-006

Table 6.1: Priorities for eco-design of products and processes.

– Use natural materials that can be recycled into biological decay cycles
– processes that release hazardous waste are avoided, and all hazardous materials are collected and reused
– for systems used to manufacture other products (e.g. machines used to produce fast-moving consumer goods such as packaged food), the criteria for the final product are taken into account
– the design must follow a set of rules [9] with a focus on basing products and processes on materials
– that can be continuously recycled without loss of performance
– that can be fully recycled into the earth's natural cycles, creating new materials
– that do not produce hazardous waste
– rely on renewable energy sources

6.2 Tool kits

Apparently industrial processes are very product-dependent. To address the problem of different products, production methods, ingredients and processes, eco-design tools have been developed [10]. A collection of tools and a manual can be found on INEDIC's website [12]. Criteria for the application of these tools have been discussed in the literature [11]. Delft University provides a website that collects tools for use in the context of eco-design and LCA [13]. Here, the MET Matrix, the Eco-Design Wheel and an example of an Eco-Design Checklist will be presented.

There are numerous publications on tools and their use in the context of eco-design – a good overview is given in [3].

6.2.1 MET matrix

The MET (Material, Energy and Toxicity Matrix, [14]) is an analytical tool that can be used to assess various environmental impacts of a product – i.e. material inputs and outputs – over its life cycle. The tool is in the form of a 3 × 3 matrix with descriptive text in each of its cells.

The columns contain the characteristic parameters for the consumption/production of waste, and the rows contain the essential life phases in a much reduced form. In more detail, this simple approach forms the basis of the life cycle assessments discussed below [4]. See Table 6.2.

As discussed in Chapter 6, consistent and detailed application results in guidelines and, where appropriate, resulting checklists that structure related audits for partners in the supply chain.

Table 6.2: MET Matrix to a composite insulating material according to [4,15].

		M Material	E Energy consumption	T Toxic materials, emission and waste
Extraction of resources and raw materials		Raw and auxiliary materials	extraction and processing	Dust particles, gases, leaching acids, pesticides, biocides . . .
Production		All materials for production and service/ maintenance	Production of intermediate and final product	Emissions into air and soil
Distribution		Packaging, diesel/ gasoline and means of transport	Packaging and packaging process	Combustion gases, fluorine, (Nano-)particles
Use	Exploitation	Materials	Electricity, fuel	
	Service	Transport and installation	Electricity/fuel	Waste in event of repair
End of life	Recovery		Electricity and fuel for transportation, recycling, waste management	Waste material after recycling, chemical substances required to access the intermediate substances
	Disposal			Remaining materials

6.2.2 Eco-design-strategy wheel (LiDS-Wheel)

One of the most popular tools in this context is the Lifecycle Design Strategies (or LiDS) Wheel, also known as the Eco-Design Strategies Wheel [16]. It is used in combination with the Eco-Design Checklist. Essentially, it is a spider web diagram that provides a multi-dimensional analysis for an intended product with the aim of comparing it to an existing (or competing) product.

It was developed under the auspices of the United Nations Environment Programme in collaboration with Delft University [17] by Hans Brezet and Carolien van Hemel – Brezet as a means of assessing how well a product design reflects the application of eight eco-design principles and is particularly useful when comparing multiple design approaches. In practical terms, each design option is represented in a spider web diagram, with the different axes representing some desirable system parameters on a subjective basis. Thus, the diagram is a useful way to visualise the various trade-offs as they are inevitably required in design. As is often the case with spider diagrams, the axes are not scaled and are therefore subject to subjective interpretation but can be used well to represent and analyse trade-offs.

Different authors suggest a different number of axes, and the guiding questions also vary [5]. The wheel provides an analytical brainstorming tool to explore areas of product development or improvement and compare the interactions between these developments, while ensuring that all phases of the life cycle are considered.

In practical use, sub-categories are proposed for each category (axle of the wheel) from which one has to choose the design approach. There are several strategy options per category as shown in the following example see Table 6.3.

Table 6.3: Criteria for the strategy wheel according to [6].

		Category strategy items
A	Innovation	1. Providing the benefits 2. Design flexibility for technological change 3. Provide product as a service 4. Serving needs that are (not) met by related products 5. Sharing among multiple users 6. Imitate biological systems 7. Use living organisms in the product system 8. Create options for local supply chains
B	Reduce Material Impacts	9. Avoidance of materials that are harmful to human or environmental health 10. Avoidance of materials that consume natural resources 11. Minimising the quantity of materials 12. Use of recycled or reclaimed materials 13. Use of renewable resources 14. Use of materials from reliable certifiers 15. Use of waste by-products
C	Manufacturing Innovation	16. Minimisation of manufacturing waste 17. Design for production quality control 18. Minimisation of energy consumption in production 19. Use of carbon neutral or renewable energy 20. Minimisation of the number of production steps 21. Minimising the number of components/materials
D	Reduced Distribution Impacts	22. Strive to eliminate toxic emissions 23. Reduce product and packaging weight 24. Reducing product and packaging volume 25. Developing reusable packaging systems 26. Use environmentally friendly transport systems 27. Sourcing or using locally produced materials and manufacturing

Table 6.3 (continued)

		Category strategy items
E	Reduced Behaviour and Use Impacts	28. Promotion of low-consumption user behaviour 29. Reduction of energy consumption during use 30. Reduction of material consumption during use 31. Reduction of water consumption during use 32. Avoidance of toxic emissions during use 33. Design of carbon-neutral or renewable energies
F	System Longevity	34. Construction for durability 35. Design for maintenance and ease of repair 36. Design for reuse and replacement of products 37. Create timeless aesthetics 38. Encourage emotional connection to the product
G	Transitional Systems	39. Design and use of upgradable products 40. Design for second life with different function Sources 41. Design for reuse of components 42. Integration of methods for collecting used products
H	Optimised End of Life	43. Version for fast manual or automated disassembly 44. Design for a recycling business model 45. Use of recyclable, non-toxic materials 46. Offering biodegradability options 47. Design for safe disposal

The following is an arbitrary example of the use of these categories in a strategy wheel to compare an existing design with a planned and actually realised design. See Figure 6.1.

6.2.3 Eco-designchecklists

Several authors offer checklists for eco-design (for an overview, see e.g. [18]). As with the strategy wheel, a set of questions is presented to help the user think about what would best meet an identified need. Below is an example of the list of questions developed by Delft University [18]. To show an example of a list that helps to apply eco-design in different steps of the life cycle. See Table 6.4 below.

Scorecards according to the lists above (often internal and standardised per product category) are often helpful if the product is very specific.

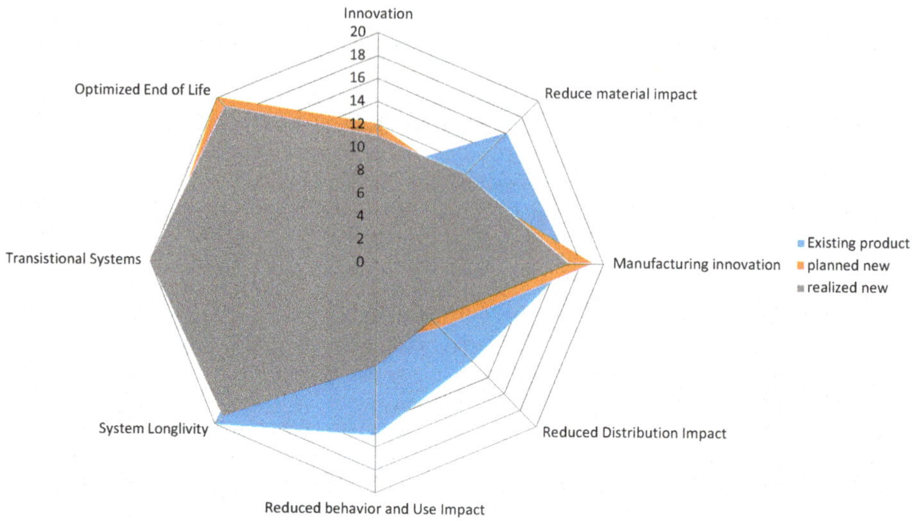

Figure 6.1: An example of using the strategy wheel to compare an existing design with a planned and actually implemented design [5].

Table 6.4: Eco-design options for different steps in the life cycle of a product [2].

Requirement analysis	
How does the product system actually meet social needs? – What are the main and secondary functions of the product? – Does the product fulfil these functions effectively and efficiently? – What user needs does the product currently meet? – Can the product functions be expanded or improved to better meet user needs? – Will these needs change over time? – Can we anticipate this through (radical) product innovation?	Eco-design strategy in the development of new concepts – Dematerialisation – Sharing of the product – Integration of functions – Functional optimisation of the product (components)

Table 6.4 (continued)

Life cycle stage 1: Production and supply of materials and components	
What problems arise in the production and supply of materials and components? – How much and what types of plastic and rubber are used? – How much and what types of additives are used? – How much and which metals are used? – How much and what other materials (glass, ceramics, etc.) are used? – How much and what type of surface treatment is used? – What is the environmental profile of the components? – How much energy is needed to transport the components and materials?	Eco-Design strategy 1: Selecting environmentally friendly materials – Clean materials – Renewable materials – Low energy materials – Recycled materials – Recyclable materials Eco-Design strategy 2: Reduction of material use – Reduction of weight – Reduction of (transport) volume
Life cycle stage 2: In-house production	
What problems can occur in the production process in your own company? – How many and what types of production processes are used (including compounds, surface treatments, printing and labelling)? – How many and what types of auxiliary materials are needed? – What is the energy demand? – How much waste is generated? – How many products do not meet the required quality standards?	Eco-Design strategy 3: Optimisation of production techniques – Alternative production techniques – Fewer production steps – lower/cleaner energy consumption – Less production waste – Fewer/cleaner production consumables
Life cycle stage 3: Distribution	
What problems can arise when distributing the product to the customer? – What type of transport packaging, bulk packaging and sales packaging is used (volume, weights, materials, reusability)? – What means of transport are used? – Is the transport organised efficiently?	Eco-Design strategy 4: Reduction of material use – Reduction of weight – Reduction of (transport) volume Eco-Design strategy 5: Optimisation of the distribution system – Less/clean/reusable packaging – Energy-efficient means of transport – Energy-efficient logistics

Table 6.4 (continued)

Life cycle stage 4: Utilisation	
What problems occur during use, operation, maintenance and repair of the product? – How much, and what kind of energy is needed, directly or indirectly? – How much and what kind of consumables are needed? – What is the technical lifetime? – How much maintenance and repairs are required? – What and how much auxiliary materials and energy are needed for operation, maintenance and repair? – Can the product be dismantled by a layman? – Are the parts that need to be replaced frequently removable? – What is the aesthetic life of the product?	Eco-Design strategy 5: Reducing impacts in the use phase – Low energy consumption – Clean energy source – Few consumables required – Clean consumables – No waste of energy and consumables Eco-Design strategy 6: Optimise initial lifetime – Reliability and durability – Easy maintenance and repair – Modular product design – Classic design – Strong product-user relationship
Life cycle stage 5: Recovery and disposal	
What problems are encountered in recycling and disposing of the product? – How is the product currently disposed of? – Are components or materials reused? – Which components could be reused? – Can the components be reassembled without damage? – Which materials are recyclable? – Are the materials identifiable? – Can they be separated quickly? – Are incompatible inks, surface treatments or stickers used? – Can hazardous components be easily separated? – Are there problems with incineration of non-recyclable product parts?	Eco-Design strategy 7: Optimisation of the end-of-life system – Reuse of products (components) – Reprocessing/refurbishment – Recycling of materials – Safe incineration

6.2.4 An approach to eco-design

There are many different approaches to eco-design – Conrad Luttropp summarised them into "10 golden rules" [19]. See Table 6.5.

These rules need to be adapted and interpreted to the specific situation if they are to be of practical use as guidelines for product development in specific situations. Rules such as the above can provide important guidance to all actors in the supply chain without restricting innovation and the introduction of new techniques.

Table 6.5: "10 golden rules" of eco-design according to [19].

1. Avoidance of toxic substances with simultaneous use of closed cycles for necessary but toxic substances
2. Minimising energy and resource consumption in production and transport through housekeeping
3. Minimising energy and resource consumption in the use phase, especially for products with the most significant environmental aspects in the use phase
4. Promoting repair options and upgrades, especially for system-dependent products.
5. Promote durability, especially for products with the most significant environmental aspects outside the use phase
6. Use structural features and high quality materials to minimise weight without compromising necessary flexibility, impact resistance or functional priorities
7. Use high quality materials, surface treatments or structural arrangements to protect products from dirt, corrosion and wear
8. Facilitate upgrading, repair and recycling through accessibility, labelling, modules, predetermined breaking points, manuals
9. Promote upgrading, repair and recycling by using few, simple, recycled, non-mixed materials and no alloys
10. Use as few fasteners as possible and use screws, adhesives, welding, snap joints, geometric interlocks, etc. according to the life cycle scenario

6.3 Eco-design – a "high-level" example

6.3.1 Introduction

As an example of value chain eco-design, the following chapter summarises a study presented to the World Economic Forum 2016 [20]. The article was aimed at a less technical audience and was intended to use the example to illustrate generally applicable requirements for production and stakeholder processes in circular economies, in particular

- No use of previously unused raw materials or combinations of materials, as in most cases this would be would not be compatible with subsequent recycling
- Labelling of all component materials to enable recycling
- Enabling easy disassembly into separate material recycling streams
- Eliminate surface treatments
- Eliminate packaging as far as possible
- Reducing weight and size for shipping

6.3.2 Approach

As an industrial system, circular economy replaces linear value chains with their "end-of-life" concept. It also shifts to the use of renewable energy, minimises the use of toxic chemicals and aims to limit waste through the superior design of materials, products, systems and – within these – business models. Often, decisions made early in the value chain (in the design phase) hinder the shift towards more circular models and material flows.

Although paper is highly recyclable, it is typically processed by various downstream industries that add through dyes and other auxiliaries. These chemicals cannot be sorted out in the sorting steps before the paper mill. In Europe alone, over 2 million tonnes of such chemicals enter paper mills every year, severely damaging the paper fibres and thus limiting their usefulness for further recycling cycles.

The recycling processes themselves cannot easily be further optimised. It is predicted that the chemicals used in the paper value chain will grow slightly faster than the paper products industry itself over the next five years.

6.3.3 Starting point

In order to provide guidance to all actors in the supply chain, it is necessary to formulate eco-design rules for paper products that do not or only slightly restrict innovation and the introduction of new technologies. To help companies also consider design and management for the circular economy, it is necessary to summarise the most important decisions of the actors and to identify the actors that can influence these decisions. In a less scientific but more descriptive way, from the perspective of a paper fibre, key principles of design for the circular economy of paper can be described:

"Don't attach anything to me that can't be stripped off or else

- 'will remain dirty and accumulate dirt';
- 'need a lot of energy to get rid of dirt';
- 'I will be discarded' ".

These principles lead to guidelines for fibre-based printing paper and packaging products but not applicable to sanitary and building paper products:

- The paper or cardboard substrate
- All materials applied to paper or packaging products to meet end-user requirements
- The processes and consumables needed to produce the paper and packaging product Paper

6.3.4 Product life cycles and circular economy

The life cycle of a paper-based product consists of a series of value-adding steps, from the extraction of resources to the end of life of the paper product and (re)recycling. In a circular economy, the end of life of the product is reconnected to its production by reusing the already extracted resources contained in used products.

Using the appropriate fibre for an application is crucial, as virgin and recycled fibres have different properties. As fibre quality deteriorates during the continued recycling process, fresh fibres are also needed in the recycling cycle. Due to the fast cycles, the amount of paper fibres available worldwide would be exhausted in about six months if fresh paper fibres were not constantly added to the cycle. The addition of fresh paper fibres starts either with the production of products that require special fibre properties or with the combination with recycled paper fibres.

In the following, a distinction is made between direct and indirect actors. Both direct and indirect actors have to take into account certain requirements in order to extend the life of paper fibres.

Direct actors in the paper life cycle "touch" the fibre in paper-based products or directly influence their quality or lifetime, while indirect actors are only suppliers of the direct actors.

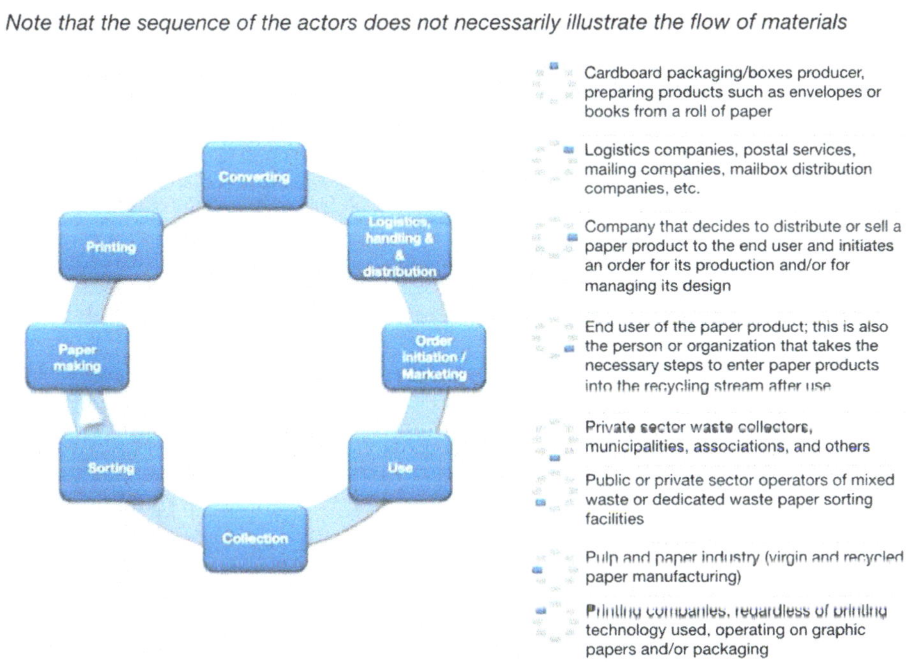

Figure 6.2: Direct actors in the life cycle of paper products.

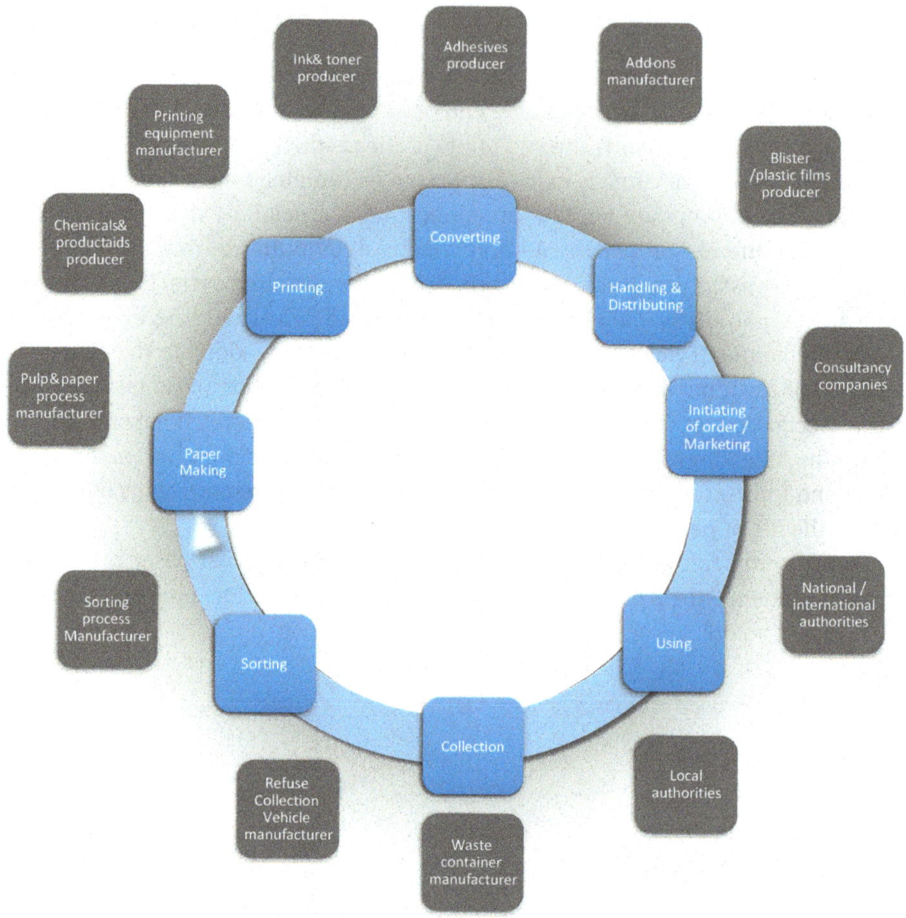

Figure 6.3: Indirect actors in the life cycle of paper products.

6.3.5 Recommendations on eco-design and management for the circular economy

A circular economy aims to maximise not only efficiency but also effectiveness. Achieving this requires awareness and knowledge of the principles of the circular economy at every stage of the value chain and a holistic approach to their application. The three core principles of the circular economy of paper products are

By following these three principles, the value of materials along the value chain is not destroyed after the product is used. In cases where current technology does not allow for the full recovery of all materials, these materials will at least not degrade or destroy the value of other materials, thus supporting the circular economy. Aspects of the recommendations relate to

Table 6.6: Three core principles of the circular economy.

– Conservation and enhancement of natural capital by controlling inventories and balancing renewable resource flows; optimisation of resource use and conservation of value along the value chain
– Optimising resource utilisation through circulation of products, components and materials and linking downstream with upstream value chains
– Promoting system effectiveness by identifying and designing out negative externalities

Table 6.7: Types of recommendations and a transversal principle to support the circular economy.

– Eco-Design: Optimising lifespan to optimise and easier recycling
– Eco-management: Managing the product to limit losses
– Environmental impact: Reducing the environmental footprint along the life cycle
– As an overarching principle: Establishing and maintaining a communication channel or platform that enables good transfer of information and collaboration between stakeholders throughout the value chain

These principles are crucial to ensure optimal material flow and high circular economy rates by avoiding leakage at the sorting and/or recycling stages.

Eco-design recommendations help to facilitate the recovery of paper fibres as the most important raw material and to preserve its properties. To this end, the key objective is to separate paper fibres from all other materials and substances used alongside the fibres to manufacture and use the paper-based product. Any element that is only difficult to separate from the fibre reduces the quality of the fibre and generates fibre losses that have to be replaced by new fibres in later life cycles. The next figure illustrates the losses (red arrows) and transfers (blue arrows) in the life cycles of paper products and shows the direct correlation with the demand for virgin fibre (green arrows). While virgin fibres are needed in the pulp and paper industry, eco-design and eco-management can help reduce resource dependence and increase value chain resilience.

The life cycle of chemicals follows similar paths: these components either end up in waste or water, or they remain in the fibre mass – and thus enter the next production cycle. The following two considerations are therefore crucial for a healthy circular economy:

- Recycled material tends to increase the concentration of substances of concern therefore the use of such substances must be limited.
- Current recycling requirements must also be taken into account in new areas of use. For example, a graphic paper could be recycled into a packaging product, which could bring additional requirements to limit residual chemical pollution.

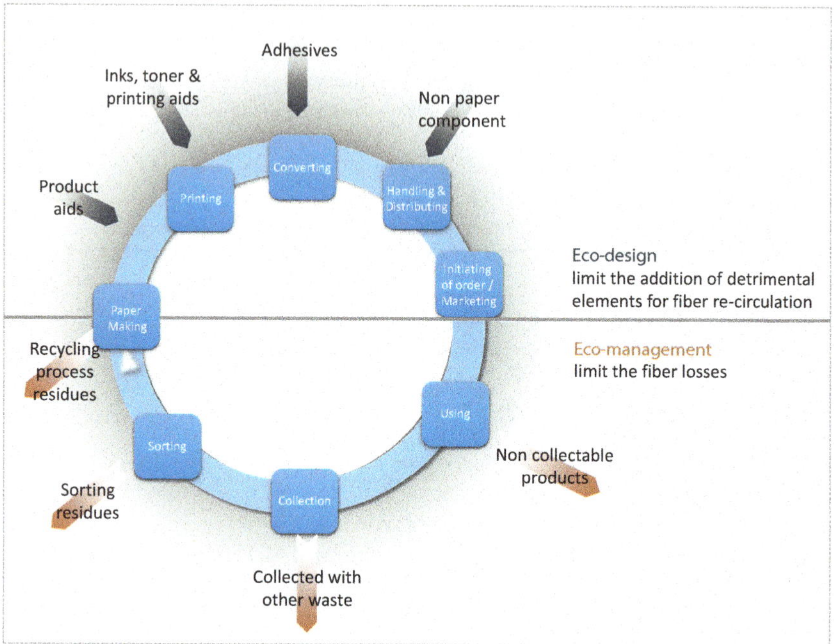

Figure 6.4: Inputs and outputs in the life cycle of paper products.

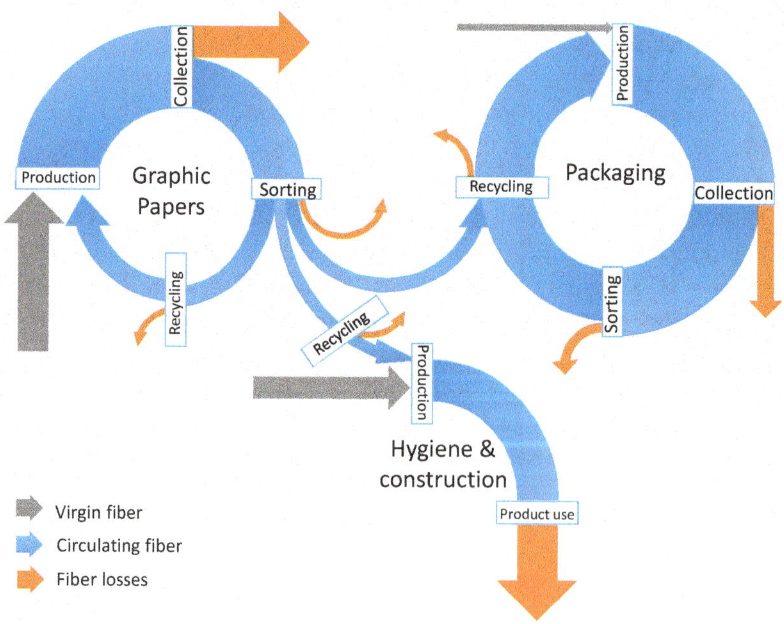

Figure 6.5: Fibre losses (red arrows) and transfers (blue arrows) in the life cycles of paper products.

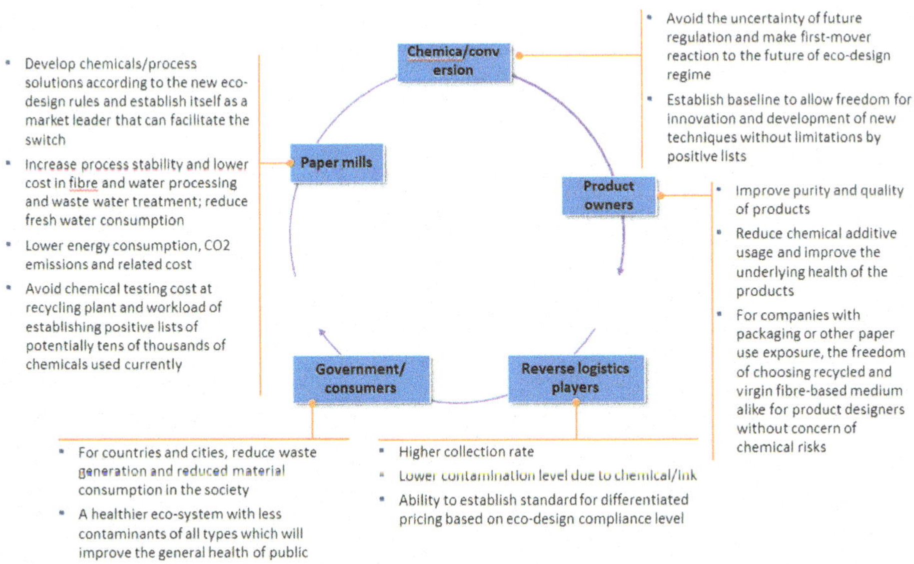

Figure 6.6: Guidelines for actors in the fibres life cycle.

6.3.6 Eco-design guidelines

In addition to the following actor-specific guidelines (table), general eco-design recommendations apply to every actor in the paper life cycle. A higher level of responsibility from all actors is beneficial for all – again formulated somewhat unscientifically from a paper fibre perspective:

“I can only be of service because I have been treated well by previous users. Please do the same so that the cycle is not broken. Minimise the amount of additives and maximise the safety of the chemicals left in me”.

6.3.7 Eco-management guidelines

Actor-specific recommendations for these guidelines should also be informed by the general principle of stewardship – any kind of degradation of properties should be minimised.

6.3.8 Recommendations for the management of environmental impacts

Figure 6.5 illustrates the main environmental impacts along the value chain of paper products. In line with the third principle (promoting system effectiveness) of the

Table 6.8: Eco-design recommendations for direct actors in the life cycle of paper.

	Type of direct actor	How to maintain my lifetime (recyclability)	How to optimize my use and quantity (resource efficiency)
Papermaking	The pulp and paper industry (virgin and recycled paper manufacturing)	If you colour me, use colourants that are easily removed with standard bleaching agents employed in paper recycling processes	Select fillers, coatings, colourants and laminates that don't disrupt my move to another cycle
		Respect my ability to keep my hydrogen bonding sites free after recycling	If I'm use to produce a recycled paper, stregthen me and enable my cleaning because I will return as raw material in your paper mill
			Use primarily recycled fibre (like me), and add virgin fibres to me that come from sustainably managed forests, in order to renew the fibre "stock" in the global flow
Converting (Changing the shape of, treating and modifying paper/cardboard)	Producers of cardboard packaging/boxes, preparing products such as envelopes or books from a roll of paper	Ensure that any other added materials can be readily separated from me	Order paper with neaar net size
		If glues/adhesives are needed to from me into the final product shape, use adhesives that can be easily removed from paper pulp, or if not, then without detrimental impact on pulp quality and waste water treatment	Recycle my scraps; sort them by homogenous grades to optimize my recycling
		Minimize adding non-paper components to me	Adjust the paper/board weight or thickness with the packaging objective (weight and other packed-product properties)
			Use closures and seals that contain no adhesives or readily removable adhesives
Printing	Printing companies, regardless of the printing technology used, operating on graphic papers and/or on packaging	Choose printing processes and materials that can be removed efficiently, and optimize the use of resources	Minimize the need to downcycle me
		If I'm a graphic paper, use ink with a good deinkability performance when printing on me (which will allow for my recycling in the graphic paper or tissue/hygienic paper industry)	Order paper/board with near net shape
		Use elements for binding (e.g. spirals, agraffes, heat-sealed bindings) that can be easily removed from paper pulp, or that have no detrimental impact on pulp quality and waste water treatment (for fold binding, for example)	Recycle my trims/scarps, and sort my scraps by nomogenous grades to optimize my recycling
		Use inks with low migration for packaging and graphic paper	Adjust the number of copies to the real needs of your target (avoid an invoicing system based on the "additional 1,000" principle); the required quantities should determine the choice of printing technology, not the opposite
		Minimize the use of ultraviolet (UV) varnishes on me, or, preferably, use UV varnishes with low impact on the recycler pulp quality (less specks)	Adjust the paper's weight with the paper product's objective
Logistics, handling & distribution	Logistics companies, postal services, mailing companies, mailbox distribution companies and others	If I'm a personalized mailing piece, use personalization printing technology with good deinkability performance	Collect and recycle the unsold or distributed products
		If samples are attached to me, choose adhesives with low impact on recycling processes (easily extracted from pulp)	Update the distribution database regularly
		Choose logistics/distribution processes that don't require extra non-fibrous materials or elements added to me that are detrimental to recycling	Simplify sorting, minimize any need for downcycling me and minimize fibre losses
Order initiation/ Marketing	Companies that decide to distribute or sell a paper product to the end user and initiate orders for their production and/or for managing their design	Clearly define the paper product's objectives before making design choices	Request a printing method compatible with the total quantity needed; avoid ordering extra copies, and optimize the product's size/format
		Design and order the production of paper product, taking into account all the above listed elements	
		Define a procurement policy for purchasing graphic paper and/or packaging	
		Define a procurement rules for advertising substrate/support (e.g. magazines, posters)	

Table 6.9: Eco-management recommendations for direct actors of the paper life cycle.

Use	End users of paper products; these are also people or organizations that take the necessary steps to enter used paper products into the recycling stream after their use	Use me in such a way that the fibre can be reused	Inform all users (households, professionals) on a global, national or local basis that all paper – graphic and packaging – is recyclable
		Avoid destroying and dirtying me	Inform all users (households, professionals) on a global, national or local basis to put me into a container dedicated to paper for recycling
		Throw me out as I am; don't crumple or shred me (except for confidential documents)	Print me on both sides where applicable
Collection	Private-sector waste collectors, municipalities, associations and others	Throw me in a recycling bin – even when you're away from home	
		Avoid collecting me waste stream with materials other than fibrous products (e.g. plastic, metal, glass)	Do not landfill or incinerate me
		Avoid collecting me in a with dirty materials and containers, and dirty refuse collection vehicles	Recycling me is preferable to composting or subjecting me to methanization (I have no specific added value for the quality of compost)
		Avoid humidity and long storage times	Avoid mixing different paper types
Sorting	Public- or private-sector operators of mixed waste or dedicated waste paper sorting facilities	Sort to grades that maximize my service/added value to the circular economy (e.g. maximize my lifetime, reduce the need for virgin fibres, emphasize cost-efficiency)	Minimize my idle time and travel distance
		Sort grades with respect to quality norms/standards	Optimize the sorting process to avoid fibre losses (effectiveness on small elements, efficiency of separation processes, adjust balling)
			Avoid humidity on stocks, where possible
			Sort grades that are "locally" requested, optimize transport distances and find a local user for me

circular economy, the recommendation is to minimise system-level impacts and negative externalities, regardless of the source, and to commit to continuous improvements in this area.

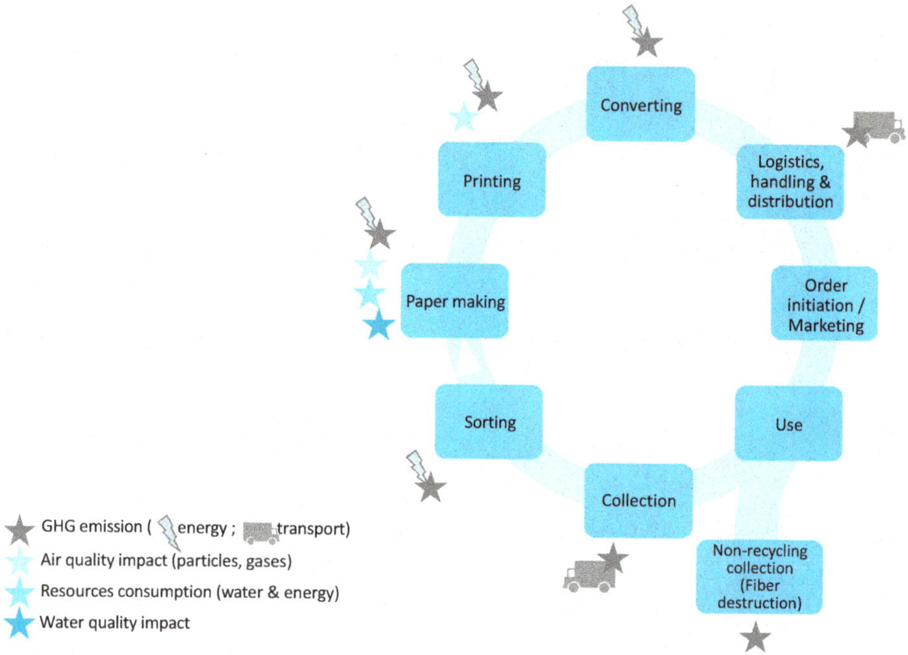

Figure 6.7: Main environmental impacts in the life cycle of paper products (the abbreviation GHG stands for Green House Gases).

6.4 Sustainable approaches to fibre-related resource management

The above approaches are crucial as resource management is a potentially significant environmental impact within the life cycle of products. In this case, forests and trees sequester and store carbon. The extraction of fibres (for pulp production) requires energy, and pulp production and paper making are water and energy intensive. If pulp is not returned to the recycling loop, it degrades, releasing greenhouse gases. Good resource, energy and waste management of pulp (Table 6.3) should therefore be an important goal for paper products along their entire value chains.

Table 6.10: Main aspects of fibre resource, energy and waste management.

1. Fiber resource management	2. Energy management	3. Waste management
1. Sustainable forest management 2. Efficient, fiber-protective pulp productin and fiber recycling 3. Consumption optimization (production of right quantities . . .)	1. Energy efficiency (process optimization, transport optimization . . .) 2. Energy mix (choice of the optimum energy mix considering the energy need and environmental impact of each type of energy . . .) 3. Local energy cooperation (combined heat and power generation, sharing energy, production capacities, identification of local energy resources . . .)	1. Quantity reduction of final disposal 2. Waste treatment choices enabling recycling, as a priority, and energy recovery secondary 3. Local industrial cooperation (my waste is your resource)

6.4.1 Order initiator supply chain checklist

End-user paper consumption is more "push" than "pull" – meaning that most paper products are sent or distributed to end-users without a specific request from them. As a result, the initiator of the order often dictates the design of the circular economy and orders and/or sells the final paper product, rather than collaborating with end-users on their consumption decisions. Therefore, a checklist specifically for the order initiator's supply chain was deemed useful. Through the order initiation process, the task of detailing some of these decisions can be delegated to other actors in the supply chain:

Table 6.11: Order originator supply chain checklist.

A. Think about and define the objectives of the product.
– What kind of information should be communicated, what product objectives should be achieved?
– What reactions can be expected from readers/consumers?
– What is the target market for the reader/end consumer?
– Has the number of copies needed been adapted to the actual needs of the target group? (Be careful with offers referring to the "extra 1,000" products). What is the lifespan of the printed paper/packaging? How long is a piece of information valid? (Or how long is it useful?)
– What image of the company is to be maintained?
– What is the exact size of the target group? How quickly can you respond to an increase in demand?

Table 6.11 (continued)

B. Identify different design choices and their implications

Deliberate choices should consider the objectives, efficiency and impact of the paper product. For example, a paper-based product must be functional; the primary purpose of the packaging is to protect and promote the product inside. Functionality is the key design determinant for the amount and type of paper used, as well as the amount and type of ancillary materials, such as inks and adhesives. Functionality is essential to maintain both the value in the packaging and the value of the materials used in the product.

C. Specification and design of the product

General

- Optimise the size of the paper/carton for the final product format to find industry standard production sizes.
- Optimise the paper weight.
- Pay special attention to the layout and artwork.
- Limit the use of coloured (tinted) graphic paper.
- Limit the addition of non-paper components and accessories and the use of wet-strength papers.
- Ensure that any non-paper product components or accessories are suitable for recycling.
- Consider the ageing of the printed paper/packaging and its contents (useful life of the printed paper/packaging).
- Match the appearance (e.g. brightness, stains, thickness) to the minimum requirements of the viewer.

Specification for non-hazardous waste in procurement

- Request to reduce or replace substances of concern.
- Preference for substances with an ecologically advantageous profile.
- For example, lower toxicity and lower persistence.
- Replace substances that are carcinogenic, mutagenic and toxic to production (CMR) with non-CMR substances, and substances that are persistent, bioaccumulative and toxic (PBT) with non-PBT substances.

Specification for the manufacture of paper products

- Promote the use of recycled paper or paper from sustainably managed forests in accordance with an internationally recognised certification scheme.
- Use printing inks that have good deinkability properties.
- Match printing technology to product life and volumes.
- Limit the use of adhesives and UV varnishes.
- Choose suppliers who are committed to environmental protection (ISO 14001 or other environmental management systems and guidelines).
- Identify and quantify substances of concern in the products as well as in the materials in the different supply chains of the final product.
- Support risk management in loops by providing information on substances of concern along the value chain(s); such information may be chemical data or risk assessments.

Table 6.11 (continued)

D. Specification for distribution and delivery – Select the most efficient distribution and logistics and delivery methods. – Indicate preference for adding less non-financial elements, e.g. add-ons (depending on the type of product). Specification for policy-makers regarding the incentive for sorting and return rate increase (different actors might be able to address their target groups in this regard). – Use national communication to inform about the recyclability of paper products. – Use local or company communication to clarify which waste stream the paper products should be disposed of in. – Communicate recyclability on packaging (if applicable); indicate that the product should be recycled after use.

References

[1] https://eur-lex.europa.eu/legal-content/EN/ALL/?uri=CELEX%3A32009L0125

[2] H. Brezet and C. van Hemel (1997) Eco-Design: A promising approach to sustainable production and consumption, UNEP, France

[3] www.INEDIC.pt, January 30, 2020

[4] Stiliyan Stefanov, APPLICATION OF "MET MATRIX" METHOD IN OUTLINING THE ENVIRONMENTAL ASPECTS OF A NEW INSULATION COMPOSITE MATERIAL, Journal of Chemical Technology and Metallurgy, 52, 5, 2017, 969–974, https://dl.uctm.edu/journal/node/j2017-5/22_16-136_Stefanov_p_969-974.pdf, Retrieved, April 11, 2020

[5] Okala – the Eco-Design strategy wheel, https://proyectaryproducir.com.ar/public_html/Seminarios_Posgrado/Material_de_referencia/OkalaEco-DesignStrategyGuide2012Final.pdf, Retrieved, April 11, 2020 http://www.engin.umich.edu/labs/EAST/me589/ecodatabasefinal/design/lids/concepts.html

[6] Sehun O, From an Eco-Design Guide to a Sustainable Design Guide: Complementing Social Aspects of Sustainable Product Design Guidelines, Sehun Oh Design, Seoul, Korea, http://aodr.org/xml/10555/10555.pdf, Retrieved, April 10, 2020

[7] http://www.wrap.org.uk/sites/files/wrap/Embedding%20sustainability%20in%20design%20%20-%20final%20v1.pdf, Retrieved April 11, 2020

[8] DIRECTIVE 2009/125/EC OF THE EUROPEAN PARLIAMENT AND OF THE COUNCIL of 21 October 2009 establishing a framework for the setting of Eco-Design requirements or energy related products, https://www.eup-network.de/background/Eco-Design-directive/, Retrieved April 11, 2020

[9] Steven Eppinger, The Fundamental Challenge of Product Design, J. Prod. Innov. 2011;28:399–40)

[10] Eco-Design Tools, http://www.ecoinvent.org/partners/Eco-Design-tools/Eco-Design-tools, Retrieved April 11, 2020

[11] A Guide for Eco Design Tools, 2nd Edition, August, 2005, http://www.fusbp.com/pdf/A%20Guide%20for%20Eco-Design%20Tools%202nd%20edition.pdf Retrieved, April 11, 2020

[12] https://www.lneg.pt/download/12235/InEDIC_MANUAL_EN.pdf, Retrieved January 30, 2020

[13] The industrial design engineering wiki, Category: Design methods and techniques, http://wikid.io.tudelft.nl/WikID/index.php/Category:Design_methods_and_techniques), Retrieved, April 11, 2020

[14] https://ec.europa.eu/environment/integration/research/newsalert/pdf/142na2_en.pdf, Retrieved, April 11, 2020

[15] Stiliyan Stefanov, APPLICATION OF "MET MATRIX" METHOD IN OUTLINING THE ENVIRONMENTAL ASPECTS OF A NEW INSULATION COMPOSITE MATERIAL, Journal of Chemical Technology and Metallurgy, 52, 5, 2017, 969–974, https://dl.uctm.edu/journal/node/j2017-5/22_16-136_Stefanov_p_969-974.pdf, Retrieved, April 11, 2020

[16] http://www.matbase.com/guidelines.html Retrieved, April 10, 2020

[17] Han Brezet, Carolien van Hemel Eco-Design: A Promising Approach to Sustainable Production and Consumption, United Nations Environment Programme, Industry and Environment, Cleaner Production, 1997928071631X, 9789280716313

[18] Eco-Design Checklist, TU Delft, http://wikid.io.tudelft.nl/WikID/index.php/Example_of_an_Eco-Design_checklist Retrieved, April 10, 2020

[19] Conrad Luttropp, 10 golden rules for Eco-Design, Stockholm, 2010, https://lpmc.lv/uploads/media/10_goldenrules_in_eco-design.pdf, Retrieved, April 10, 2020

[20] M. Has et al, Design and Management for Circularity – the Case of Paper; World Economic Forum Davos 2016; http://www.cepi.org/publication/design-and-management-circularity-%E2%80%93-case-paper, Retrieved April 11, 2020

Chapter 7
Life cycle assessment of an organisation

7.1 Introduction

Generally speaking, in terms of sustainability aspects, a company can be divided into an organisation and the products it produces. While the product-related aspects are dealt with above, the environmental footprint of an organisation is subject to a somewhat different approach than a product-related analysis. To this end, the EU has published an "Organisation Eco-footprint (OEF) Guide" [1], which provides the necessary information to conduct an OEF study. The guide has a strong methodological character and is structured in phases that provide assessment approaches, best practices, tips and hints with examples and ways to interpret and classify the results. The contents of an organisational environmental footprint and ways to review it are offered. Other methodological guides and standards to consider regarding an organisation's environmental footprint are given in Table 7.1.

Table 7.1: Standards to be considered in determining the environmental footprint.

- ISO 14064 (2006): Greenhouse gases – Parts 1 and 3
- ISO/WD TR 14069 (working draft, 2010): GHG – Quantification and reporting of greenhouse gas emissions for organisations
- The ILCD (International Reference Life Cycle Data System) Handbook (2011)
- The Corporate Accounting and Reporting Standard of the Greenhouse Gas Protocol (WRI/WBCSD) (2011a)
- DEFRA – Guidance on how to measure and report our greenhouse gas emissions (2009)
- The Carbon Disclosure Project for Water (2010)

The purpose of this chapter is to provide an overview of the considerations behind establishing an organisation's environmental footprint. A particular focus is on the identification and involvement of relevant stakeholders. In day-to-day operations, the commitment and involvement of all stakeholders is crucial to the success of the endeavour. Reasonable efforts must be made to reach the widest possible consensus during the process (adapted from ISO 14020:2000, 4.9.1, Principle 8). It is often helpful to involve experienced consultants – also to achieve comparability of the results achieved.

In general, the information provided for this comparison must be transparent, especially so that the user can understand the approaches and limitations and, if necessary, address them in a targeted manner.

https://doi.org/10.1515/9783110767308-007

7.2 General approach

Since it is not easy to allocate consumption within a company to a specific product, it makes practical sense for a company or site to separate site-related/general administration, procurement, marketing, service and product-related activities such as production. The first step to determine the environmental footprint must be an accounting with a clear allocation to activities, for example, with a matrix such as the one given in Table 7.2.

Table 7.2: High-level matrix for summarising organisational and product-related aspects of the contributions to an environmental footprint.

Emissions/ materials	**Materials**	**Site**					**Relation to products/ portfolio Which product?**
		Admin	Procurement logistics	Marketing	Service	...	
Scope 1							
Scope 2							
Scope 3							
Materials in	Specify ...						
	Water						
	Poison						
Material out	Specify ...						
	Water						
	Poison						

In the case of product-based organisations, the location-based aspects must be assigned to the products produced at the end.

7.2.1 Priority principle

In most companies, rudimentary approaches like this kind of matrix already exist. For an organised approach to the product aspect of a project, the "flow" has been described in the literature (e.g. [2]; Table 7.3).

Table 7.3: Action flow for the assessment of environmental footprints.

A) Definition of goal and scope: The first step involves defining the goal and scope by understanding the requirements. This involves mapping and engaging stakeholders to identify and get their agreement on the example boundaries, attributes and impact categories, identifying the ground rules of data mining and analysis and the need to use proxy data where required. Essentially, this step involves defining the objective, scope and agreeing on a prioritisation approach for the following example
B) Data collection, expertise and analysis: In addition to LCA, this step includes data collection, data analysis, validation, expert interviews and stakeholder consultations, and the synthesis of different evidence, such as life cycle study results, input/output analysis data, scientific research studies, product information, sales volumes/economic value and trade information, depending on the scale and scope of the study (see, e.g. [3]).
C) Identification and validation of hotspots: Once all the necessary data and knowledge are collected, the criteria for identifying, ranking and prioritising hotspots are developed, discussed and validated. This includes identification and prioritisation of opportunities to reduce impacts, review and validation of the identified hotspots by a wider group of people, and identification of implementation gaps and recommendations to reduce impacts.
D) Prioritisation of actions: The final step involves using the results of the study to make sustainability improvements. Typically, this step includes action planning, development of industry guidance and standards, piloting or field testing of potential solutions, industry collaborations and voluntary agreements, etc., and further engagement with relevant stakeholders to disseminate and mainstream proven or effective solutions based on feedback from pilot activities.

For an organisation, the EU report [1] offers the sequence of actions given in Figure 7.1.

7.3 Sustainability of an organisation

Entrepreneurial or organisational activities, even if not related to a physical product, result in an environmental footprint. This is independent of their relation to projects or relates only to internal practices.

Means have therefore been developed to measure and reduce the environmental impact of an organisation [4]. In essence, the approach is very similar to what the EU has proposed [1].

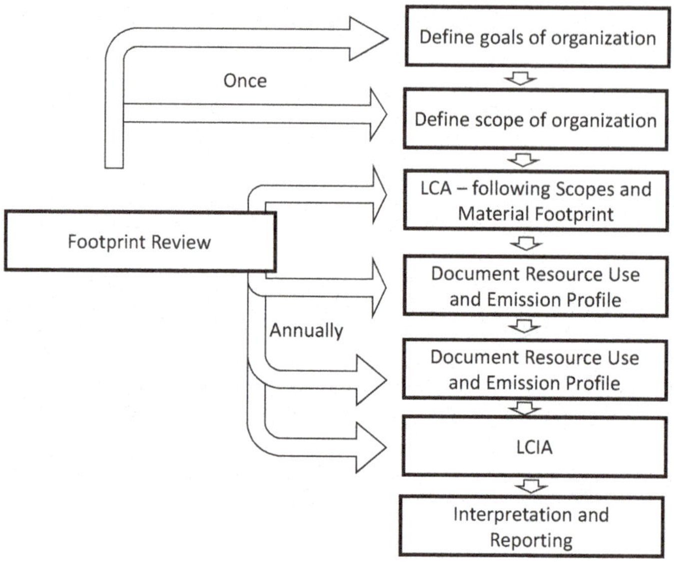

Figure 7.1: Sequence of activities for creating an organisational environmental footprint (based on [1]).

7.3.1 Background

Offices, like other organisational units, need to use resources and therefore have an environmental footprint. Resources are typically related to the provision of water and energy (electricity or even gas) as well as an infrastructure for the disposal of waste (waste water, paper, food or food packaging, etc.) or even the emission of greenhouse gases from travelling and transporting goods. Even if this can be considered or is in fact less impactful than the environmental footprint created by the production of physical products, the environmental impacts associated with such activities can be quantified, analysed and, if necessary, reduced. Structurally, and for all types of organisations, the issues to consider when looking at their environmental footprint are similar (Table 7.4).

Table 7.4: Key parameters when considering organisational environmental footprint.

– Energy use within buildings – such as electricity for lighting, running computers, heating or cooling systems, and cooking
– Transport – both for work purposes and for employees to travel to work
– The consumption of goods such as equipment, stationery and food
– Consumption of water
– Production and processing of waste
– Use of buildings and infrastructure in buildings, including cooling and insulation

The above parameters for the environmental footprint are general in nature – therefore require adaptation to local conditions. To motivate also personal actions, their expected benefits as well as concrete instructions for action should be in focus – worth mentioning are, for example, those presented in Table 7.5.

Table 7.5: Benefits of good environmental stewardship according to [4].

Limiting use	Conservation of resources	Limiting the use of paper, plastics, metal and water to reduce depletion of natural resources.
Save	Cost savings	Save expenses by efficient use of resources (such as turning lights off when there is no one in the room).
Save Health	Health improvements	Well-insulated buildings stay cooler in high temperatures and warmer in colder temperatures, reducing the required energy consumption for heating or cooling. Staff who work in well-ventilated offices are likely to be more productive and stay healthy.
Efficiency	Improved projects	Plan projects so that minimal impact on and from the environment is achieved. The effectiveness and sustainability of projects are therefore improved.
Reduce footprint	Reduced CO_2 emissions	Reducing the use of fossil fuels through reduced energy consumption and increased efficiency in transport and commutes.
Reliability	Enhanced reputation	Better stewardship of resources improves the organisation's credibility among all partners and the community who view the organisation as caring and responsible. Organisations may also experience an increase in financial support as a result.
Risk reduction	Meet national guidelines and laws	National governments are held accountable for their actions by international agreements, particularly in relation to climate change and carbon emissions. As discussed in Chapter 1, governments are and will be putting more pressure on industry, the public sector and citizens to contribute to the achievement of national targets.

In addition, it is important to emphasise the personal component in the catalogue of measures, that is, the reduction of waste and unnecessary consumption of resources through, for example, personal allocation of consumed resources, which in effect amounts to holding employees accountable, sensitising them and training them. Employees need to be and remain reminded and motivated. The implementation of appropriate policies must be monitored and regularly reviewed against targets set in an environmental audit. It is obviously important here that visible leaders set a good example.

7.3.2 Best practices

One of the useful approaches to good environmental management is to refer to best practices. Once implemented, these practices can be used to benchmark and maintain quality [6]. For offices and organisations, the following practices have been found to be positive (Table 7.6).

Table 7.6: Established practices for reducing the environmental footprint of organisations. Corresponding efforts for office-related environmental protection should be worded in an environmental strategy.

	Best practice
Reducing the use of energy and associated greenhouse gas emissions	Sourcing goods and services locally wherever possible
	Ensuring that all electrical equipment is turned off when not in use such as not left on standby (which still uses electricity) and switching off lights, air-conditioning, fans and heating when not needed
	Using energy-efficient light bulbs
	Opening windows or having meetings outside rather than using air conditioning, when possible
	Sourcing of energy via renewable resources
Reducing staff travel	Co-ordinating visits by different staff members to project sites to reduce the number of trips and distance travelled, and to enable staff to travel together
	Sharing vehicles or combining travel with other NGOs to adjacent sites
	Walking or using public transport as much as possible
	Making more use of low fuel transport when possible – such as motorbikes rather than four-wheel drive vehicles
	Reducing the number of flights and using phones, internet technology and video conferencing, wherever possible, as alternatives to travelling to meetings
Reducing the use of other resources	Avoiding unnecessary printing out of emails or documents unnecessarily such as using both sides of paper
	Re-using envelopes and packaging Encouraging electronic correspondence and file sharing

Table 7.6 (continued)

	Best practice
	Reducing waste and pollution through: – using biodegradable chemicals where possible such as separating rubbish to allow composting and recycling (provide separate bins), – banning the use of plastic bags.
Other initiatives	Even if there may not be the opportunity to recycle materials, consider to donate paper, card board and glass to local initiatives as that may be of value after collecting and selling them. Organise regular collection points for them.
	Collection of wasted but still useful materials such as kitchen and food waste for poorer people and animal feed compost. Consider establishing a recycling project.
	Energy-saving strategies including investing in renewable energy supplies, upgrade insulation or natural cooling, and fitting mosquito netting over windows so natural ventilation can be used.
	Ethical procurement to reduce transport-related emissions such as choosing local products and materials over imported ones. Sourcing products only from companies that are seeking to reduce their footprints and encouraging contractors to take account of the organisation's environmental policy when carrying out work, such as using timber from sustainably managed forests.
	Improved water management such as installing rainwater harvesting tanks to reduce freshwater consumption.
	Consider to provide less meat in canteens.

7.3.3 Environmental audits for organisations

The need for organisations to conduct a regular environmental audit should be included in the environmental policy. An audit measures the organisation's performance in reducing its negative environmental impacts and involves the collection of accurate, comprehensive and meaningful information. The initial audit can and should be used to establish a baseline against which progress can be measured.

An audit can be carried out by staff or by an independent auditor. Alternatively, two organisations can agree to audit each other in a "peer review" manner and then review the results together, which promotes the learning process.

For many organisations, the end of the financial year is the best time to conduct an audit, when other records are being updated. They conduct an environmental audit when other records are being updated. Most measurements are best taken on an annual basis so that seasonal patterns such as weather patterns or holiday periods do not result in widely varying measurements. Many of the measurements should be fairly easy to collect, such as noting electricity consumption, recording kilometres travelled by employees, business trips or checking the amount of paper used or even various of the measures mentioned above. A monthly accounting system facilitates the audit.

Employee surveys may need to be included as part of the audit, for example, to find out how employees travel to work. Where measurements are missing, an estimate can be made based on measurements from similar organisations, if available. New recording practices can then be established for the following year. It is important to use consistent measurement methods between one audit and the next so that progress can be monitored accurately.

A meeting should be held with all staff at the beginning of an audit to clarify audit activities and requirements, motivate and both outline what is expected and ensure that the audit is perceived positively.

The following table contains general parameters for measurable or countable consumption or waste. Calculating the totals obviously requires careful and planned collection and verification of information, especially if there are many employees. Usually the first audit is more time-consuming than the following ones and the cooperation of external experts is recommended – therefore the first audit is more often cost-intensive than later ones, which are then also less time-consuming if the recording of the necessary data becomes part of the office routine.

For each measurement, an audit should set a target for the following year. At the end of an audit, the measurements should be compared with those of the previous year to see if improvements have been made and the targets achieved. Obviously, efforts need to be made to understand deviations and address areas where targets have not been met.

Subsequently, the same structure is applied as used in the above sections based on the scope questions (Table 7.2).

After completing the environmental audit, it is recommended to identify the largest sources of waste and emissions and to develop and implement a policy with an action plan to reduce them year by year. It may be helpful to share audit information to make some comparisons with other similar organisations or comparable departments in nearby companies. This way, organisations can check how well they are doing and exchanging useful ideas.

The results of the audit must, as always, be well documented, for example, in a report. This document should include the contents given in Table 7.7.

Table 7.7: Minimum content of the environmental audit results report.

- Description of the contents and methods of the audit (which bodies and persons, the type of organisation, details of the issues addressed, results and steps for further action)
- Qualitative and quantitative presentation and evaluation of the results
- Information gaps and estimates, literature references
- Identification and naming of good and bad environmental practices
- Proposals for action and targets

In addition, the report should describe targets for all tasks and to what extent in particular (Table 7.8).

Table 7.8: Targets achieved to be reported in environmental audit results report.

- Compliance with government regulations, guidelines, codes of conduct and permit conditions was achieved
- Where there is potential for improvement separated into regulated and non-regulated areas with the aim of minimising footprints
- Potential savings are used and explored
- Additional general objectives should be pursued
- The activity is integrated into the general management system

References

Sources 7.1.2

[1] https://www.esg.adec-innovations.com/about-us/faqs/what-is-ghg/

[2] www.climatechange.vic.gov.au/legislation/climate-change-act-2017

[3] https://ec.europa.eu/commission/presscorner/detail/en/mex_20_2389

[4] https://unfccc.int/process-and-meetings/transparency-and-reporting/reporting-and-review-under-the-convention/greenhouse-gas-inventories-annex-i-parties/national-inventory-submissions-2019

Sources 7.1.3

[1] Muthu, Subramanian Senthilkannan, Assessment of Carbon Footprint in Different Industrial Sectors, Volume 1, 2013, Springer, SN – 978-981-4560-40-5

[2] GHG Protocol, www.ghgprotocol.org/standards

[3] https://reports.climatecentral.org/pulp-fiction/1/

Sources 7.1.5.1

General literature: Leaflet on dealing with sustainability Risiken BAFIN Annual Report (bafin.de)

[1] https://www.eciia.eu/2021/01/risk-in-focus-2021-practical-guidance-on-climate-change-and-environmental-sustainability/

[2] https://www.feri-institut.de/glossar/t/transitionsrisiken/

Sources 7.1.5.3

[1] Climate protection in figures: CO2 pricing (https://www.bmu.de/publikation/climate-action-in-figures-2019/)

[2] Basis for CO2 price stands (https://www.bundesregierung.de/breg-de/themen/klimaschutz/nationaler-emissionshandel-1684508#:~:text=Bund%20und%20Länder%20einigten%20sich,und%20höchstens%2065%20Euro%20gelten.

[3] ETS: Price for CO2 certificates exceeds 32 euros (https://www.energiezukunft.eu/wirtschaft/preis-fuer-co2-zertifikate-erreicht-allzeithoch/)

[4] Basis for CO2 price stands (www.finanzen.net/rohstoffe/co2-emissionsrechte)

[5] Indicator: Greenhouse gas emissions (https://bmu.de/en/indicator-greenhouse-gas-emissions)

[6] Federal Agency for Civic Education From 2021: CO2 price on heating and motor fuels https://www.bpb.de/politik/hintergrund-aktuell/324668/co2-preis-auf-heiz-und-kraftstoffe

[7] Carbon Price Viewer – Ember (https://ember-climate.orgdata/carbon-price-viewer/)

[8] Global CO2 emissions: How 2020 could become a turning point in the fight against climate change (https://www.bbc.com/news/science-environment-55498657)

[9] Greenhouse gas emissions in Germany | Federal Environment Agency of Germany (https://www.umweltbundesamt.de/en/indicator-greenhouse-gas-emissions#at-a-glance)

[10] Internet Record decline in global CO2 emissions – Max Plank Gesellschaft (https://www.mpg.de/16175501/1214-ebio-corona-pandemie-fuehrt-zu-einem-rekordrueckgang-der-globalen-co2-emissionen-152860-x)

[11] Internet NATIONAL CLIMATE TARGETS 2021-2030 FOR NON-ETS SECTORS (https://ec.europa.eu/clima/policies/effort/regulation_en)

Chapter 7.2

[1] http://lct.jrc.ec.europa.eu/pdf-directory/ILCD-Handbook-General-guide-for-LCA-DETAIL-online-12March2010.pdf/view

[2] EU Commission, Organisation Umweltfußabdruck (OEF) Guide, https://ec.europa.eu/environment/eussd/pdf/footprint/OEF%20Guide_final_July%202012_clean%20version.pdf, Retrieved April 11, 2020

[3] DIRECTIVE 2006/42/EC OF THE EUROPEAN PARLIAMENT AND OF THE COUNCIL of 17 May 2006 on machinery, and amending Directive 95/16/EC, https://eur–ex.europa.eu/LexUriServ/LexUriServ.do?uri=OJ:L:2006:157:0024:0086:EN:PDF, Retrieved April 11 2020

[4] Kirchherr, J., Reike, D. & Hekkert, M. Conceptualizing the circular economy: An analysis of 114 definitions. Resour. Conserv. Recycl. 127, 221–232 (2017)

[5] Energy-Using Product Group Analysis – Lot 5 Machine tools and related machinery, Fraunhofer Institute for Reliability and micro integration, IZM Department Environmental and Reliability Engineering Dipl.-Ing. Karsten Schischke, Dipl.-Ing. Karsten Schischke, Berlin, August, 2012

[6] Nash, Jennifer; Ehrenfeld, John (2010). "Code Green: Business Adopts Voluntary Environmental Standards". Environment: Science and Policy for Sustainable Development. 38: 16. doi:10.1080/00139157.1996.9930973

Chapter 8
Energy and material strategy

In the context of reducing the environmental impact of a product or organisation, strategy refers to the reduction of materials, resources or energy. Measuring the amounts of materials, energy or emissions consumed, as described above, is a necessary prerequisite for implementing this strategy.

The idea is to,

- use less material and energy to achieve the same or more performance (less space, more function, higher utilisation, sharing) over a longer lifetime
- produce less weight
- allow reuse of components
- allow and promote recycling, upcycling
- allow improved yields in all processes
- use less and renewable energy in all phases of use

There are interactions between implementation mechanisms, especially when trade-offs between material and energy efficiency become necessary: The pathways that lead to downsizing or more intensive use often exploit synergies between material and energy efficiency, while others, such as lifetime extension or weight reduction have trade-offs in terms of durability. Against this background, broader system boundaries need to be considered in the analysis and product definition. These broader system boundaries also mean that savings are often not as great as originally assumed.

Synergies and trade-offs between material and energy efficiency need to be considered in the selection of strategies and subsequent design of measures.

8.1 Material strategy

8.1.1 Accounting and return logistics

As seen above, the environmental footprint is influenced by consumption of both energy and material – whether as absolute mass contained in the product or as waste. Consequently, strategies need to be developed to reduce processes, product design in terms of consumed mass of materials used during all phases of the product life. This requires a collective organisational effort, especially by R&D and manufacturing, based on an integrated strategy that should be co-developed and applied in the early design phase of the product. Two standards on material efficiency are worth considering

- ISO 22628 – Road vehicles: recyclability and recoverability, calculation method; and

https://doi.org/10.1515/9783110767308-008

- IEC 62635 – Guidelines for end-of-life information provision from manufacturers and recyclers, and for recyclability rate calculation of electrical and electronic equipment.

Source [1] provides an overview of the literature on resource and material efficiency indicators, including an initial assessment of their suitability for detailed analysis in the project. What can be influenced in the early product design phase is sketched in Table 8.1.

Table 8.1: Parameters to be influenced in the early product eco-design phase.

- Material accounting and allocation of reasons for use
- Avoidance of non-recyclable materials, non-reusable parts
- Reduction of material diversity
- Substitution of harmful and hazardous substances and materials
- Reduction of material consumption in production and use
- The ability of the system to use recycled and recyclable materials, including biodegradable materials
- Enabling reuse and refurbishment as parts or as raw materials
- Avoiding composite materials and variations (of alloys) to increase the value and recycling rate of product parts
- Applying special design measures plus service to extend the life (of the system itself and the parts)
- Extending the life of parts and enabling a second life after overhauls.
- Changing processes, which may imply changing partners
  in addition
- Ways to simplify the assembly and disassembly of parts and products
- Enabling and facilitating repair, reuse, take-back, recovery, recycling
- Selection of materials and consumables, appropriate packaging of materials
- Management of machines in the field and their return for disassembly
- Legal access to returned materials

In practice, it is often helpful to define KPIs or priorities beyond the measures that already exist by law for

- Use of "low impact" materials (e.g. recycled/recyclable/renewable/certified as sustainably sourced), for example by linking to waste minimisation targets – for example reducing material weight of products
- Green chemistry – product safety/no toxic components or processes – link to targets on product sustainability attributes.
- Use of raw materials and ingredients that help reduce the environmental impact of the product or realise new environmental benefits Link to targets on supply chain environmental footprint
- Reduced environmental footprint

Source [1] provides an example analysis of material efficiency parameters in existing guidelines and studies as seen in Table 8.2:

Table 8.2: Analysis of material efficiency parameters and units (according to [1]).

Aspect	Parameter	Unit	Related to product design		Quantifiable parameter
			Product and components	Component only	
Quantity of material used over life cycle	Weight of product	kg			
	Consumption of materials (over the life cycle)	kg			
	Material footprint	kg			
	Material inputs per unit service (MIPS)	kg per unit of service			
	Material intensity (DMC/GDP)	kg/€			
	Solid waste intensity (t/GDP)	t/€			
	Weight utility ratio	kg/kg			
	Consumables needed	kg, l, or unit			
	Availability of take-back systems for consumables	None			

(continued)

Table 8.2 (continued)

Aspect	Parameter	Unit	Related to product design		Quantifiable parameter
			Product and components	Component only	
Environmental impacts of extraction, production and end-of-life of materials	Abiotic depletion potential: mineral, fossil	(kg Sb equivalent)			
	Quantity of hazardous substances[72], quantity of heavy metals[73] (in the product)	g, g/kg or %			Quantity of hazardous substances[72], quantity of heavy metals[73] (in the product)
	Critical dilution volume[74]	g/g			Critical dilution volume[74]
	Biodegradability	g, g/kg or % per period of time			Biodegradability
	Waste generated (over the life cycle)	kg			Waste generated (over the life cycle)
	Recyclability benefit rate (of product)	%			Recyclability benefit rate (of product)
	Recycling rate (waste statistics)	%			
	Recoverability of	Separability and easiness of			

Recoverability of materials/product	Separability and easiness of disassembly	None	
	Time of disassembly	S	
	Recoverability benefit rate (of product)	%	
	Recovery rate (waste statistics)	%	
	Marking of materials/components	None	
Origin of materials	Recycled content	% in weight	
	Re-used components	% in weight	
Reusability of components/product	Raw materials with sustainable origin	(% in value or % in weight)	
	Reuse rate (waste statistics)	%	
	Reusability benefit rate (of product)	%	
Reparability and durability of components/product	Performance and durability parameters after a certain amount of time	Specific performance parameter of the product	
	Warranty time	year	
	Availability of spare parts	None	
	Ease of repair and upgrade	None	None
	Lifetime	year	✓

For accounting purposes (also in [1]), an overarching list of materials (Table 8.3) to be preferred for eco-design has been proposed. It should be noted that the list needs to be extended or reduced depending on the type of product considered.

Table 8.3: List of materials used.

Type of material		Parts number	Footprints, Water, materials consumed along supply chain	Weight	Use for reuse, repair, recollection, recovery, recycling
Metals	Chemical elements and Alloys				
Non-metallic minerals					
Composite and bonded materials (not preferred)					
Essential process materials	Polyethylene (PE)				
	Polypropylene (PP)				
	Polyethylene terephthalate (PET)				
	Glass				
	..				
Bio-materials	Wood				
	Paper and carton board				
	..				

As mentioned planning for the reuse of materials and systems is by no means trivial. It requires that the system or parts in question be returned to the supplier, for which, in addition to a verified business model, well-established logistics are crucial [2]. However, logistics is usually a more complex obstacle, and ideally the system is installed and operated close to the production site and the central service and maintenance centres. In practice, this ideal situation often does not exist. For service and inbound logistics, this is not a problem, as long as the organisation is suitable for it.

The returns policy (outbound logistics) is more complex because

- the customer must agree in advance to the return of the system
- the user/installation characteristics must be taken into account
- and a collection and sorting process must be in place to distinguish parts that are to be reused as is, tested and repaired, refurbished or recycled to serve as a primary resource [2].

Industrialised dismantling with appropriate reuse of parts can only start if enough parts of suitable quality come back from the field. The number of machines in the field must obviously be sufficient for a plannable period and continuous material flow. If this and a suitable constructive preliminary work are given, concepts for local reflow, testing and reuse can usually be created.

Even if the energy expenditure for transport is lower than the gained equivalent of material and energy for reprocessing, the question remains whether it is worth the effort to set up an organisation and the corresponding logistics to implement the take-back. There is no single answer to this question that applies to all types of systems and regions concerned.

As indicated above, the implementation of a return policy implies that the customer agrees to return the system after the use phase when signing the contract. Implicitly, this challenges the concept of ownership of what is purchased, as established in some markets. Although the concept of circular economy is not yet established for industrial goods, legislation requires customers to commit to returning products for refurbishment where possible once the product has reached the end of its life.

8.1.2 Material recycling as a business model – an example

The following is a practical example of a possible approach to a take-back, disassembly and re-use strategy [3]. The reference case is an international printing equipment supplier that focuses on the development and distribution of B2B and B2C products. Most of the time, a service contract is part of this company's business to ensure customer satisfaction, as they primarily value reliable uptime. To this end, a fast recovery of the machine in case of service is crucial. Systems are also developed for this with a defined spare parts philosophy – fast replacement only for what is needed. With this in mind, a logistics system exists that is suitable for remanufacturing. Remanufacturing has been seen as part of the remanufacturing of equipment for some time. The approach is internally referred to as circular economy manufacturing. In addition to reducing integral costs, the task of the corresponding team is to reuse parts, modules and entire systems or just the materials. At the same time, design for disassembly or design for reuse is promoted.

The underlying strategy is embedded in a "Design for Dismantling" and "Design for Reuse" initiative that is developed and implemented during the early product design of the respective products. Experience shows that this type of product design strategy must be set as a goal from the beginning and implemented before the product specification stage. This strategy or its features should not be influenced or changed during or after industrialisation. It must be finalised and implemented after the discovery phase of the R&D process and before industrialisation.

Elements of the strategy to be considered are

Table 8.4: Points to consider for implementing a material strategy in the design phase according to [3].

1. Ensure simple disassembly techniques.
   (e.g. no adhesives or composite materials)
2. Ensure simple disassembly structures.
   (number of parts, access and tools)
3. Think of multiple life cycles in the design phase.
   (stable frames, easily repairable units)
4. Use reusable materials.
   (easy to clean, easy to reuse)
5. Think in platforms.
   (design not for a product but for a platform)
6. Design parts for reuse.
7. Ensure that wear parts are easily accessible without dismantling the whole unit.
8. Especially avoid low-quality stickers as they are difficult to remove.
9. Develop smooth surfaces that are easy to clean.
10. Do not change the design unless it is really necessary.
11. Avoid paint.
12. Avoid grease.
13. Avoid tight joints.

Permanent connections in particular pose a problem for disassembly – effort and complexity are often an obstacle. The following figure 8.1 illustrates how the avoidance of permanent connections is implemented to achieve a good reuse design.

Internally, the company has a variety of R&D standards for disassembly that are implemented during the design phase, as well as work-related instructions based on R&D instructions. Parts remanufacturing is applied after disassembly and environmentally sound return of used machinery and disassembly based on disassembly instructions.

The process before accepting a part as one of the qualified Second Life spare parts is organised in sequential steps and is of course very product dependent, but profitable for all products launched so far is sketched in Table 8.5.

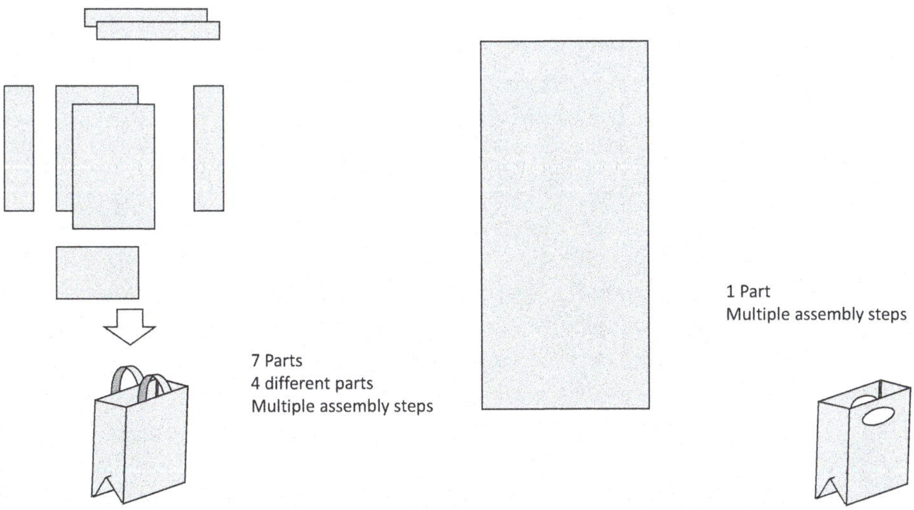

Figure 8.1: Avoiding a permanent connection to facilitate good reuse according to (based on 3).

Table 8.5: Process before accepting a part as one for qualification of second life spare parts [3].

1. Identify the opportunity, build the initial business case.
2. Collect parts for pre-production, create test document.
3. Decide on internal or external repair.
4. Review specifications from R&D.
5. Create instruction for reuse (e.g. upgrades/replace parts/test/clean/ package).
6. Run pilot series and test, finalise reuse instruction.
7. Review business case.
8. Release quality department.
9. Regular production.

The role of the respective department is ultimately multiple – where the department must be commercially profitable:

- Supplier of defined parts for specific systems (field service and manufacturing supplier).
- Supporting the planning and design of equipment and modules, as well as the country organisation in the return of equipment.
- The development of equipment-dependent business models.
- Dismantling of systems.
- Delivery of segregated waste to suppliers.
- In sum: reduction of the carbon footprint of a product.

8.2 Energy-related strategy

So far, only a combination of two means of enforcing a reduction in energy consumption, measured in CO_2 emissions, has been successful:

1. Carbon emissions tax or
2. Cap-and-trade systems.

Both systems also exist simultaneously within one legal system. For example, the CO_2 tax on cars, as levied since 2021, is an emissions tax. The concept of a cap-and-trade has been adopted, for example, by the European Union European Emission Trading Scheme. The EU introduced it in 2005 to meet its cap set in the Kyoto Protocol.

While the first scheme aims to reduce emissions by incentivising individuals and businesses to become more efficient and use cleaner energy sources at reduced rates, the second sets an overall limit or cap on emissions. In that case, the right to emit carbon is then allocated or auctioned and these emission rights can then be bought and sold on the carbon market. In both cases, there is an incentive to reduce emissions. As discussed earlier, it is necessary to establish a starting point and subsequently carry out continuous and regular monitoring, reporting on the targets achieved and not achieved in each accounting item.

An energy strategy is based on the need to reduce the costs associated with emissions as well as other costs, while operating in a future-proof business manner.

The system for monitoring and reporting energy consumption and GHG emissions must be stable, accurate, transparent and in compliance with the laws and legal standards applicable in the sector and region under consideration. As already mentioned in the discussion on material strategies, all measures start with accounting and the correct allocation of consumption quantities. The structure of the examples from the above chapters (according to the scopes of the GHG) is sufficient for this purpose. Per sub-criterion of each scope, in order for the impact of the implemented measures to lead to the desired results, strategy, targets, reporting and monitoring should be addressed.

Many aspects of energy-related strategies and reports have already been addressed in the examples described above and in the audits and their reporting (Chapter 8.1). A report should also include the following information contained in Table 8.6

Simple measures to reduce the footprint include reducing unnecessary energy consumption in equipment and offices (the use of smart sockets, shutdown and downtime, the use of flexible and remote workstations to reduce energy consumption in the office), in production one can focus on the general reduction of energy consumption in the production and use of the produced goods, with a focus on the use of renewable energy sources.

Table 8.6: Minimum content of an energy footprint report.

- Main characteristics of the company (see e.g. questionnaire above).
- Review of the year of analysis and the previous year(s)
- Complete set of data on emissions (according to the scopes equivalents of carbon dioxide t CO_2 eq., (see e.g. [4]) from local site(s), supply chain(s) and effort for waste/degradation/recycling) – see also the examples above
- Responsible persons (names and area of responsibility)
- Measures/training attended
- Documentation on plan and target achievement, data collection and calculation methods used, unaccounted emissions
- Audits and assessments carried out

References

[1] Material-efficiency Eco-Design Report and Module to the Methodology for the Eco-Design of Energy-related Products (MEErP), PART 1: MATERIAL EFFICIENCY FOR ECO-DESIGN, Final report to the European Commission – DG Enterprise and Industry, 5 December 2013, 65.

[2] Kuo, Tsai. Simulation of purchase or rental decision-making based on product service system. International Journal of Advanced Manufacturing Technology. 52. 1239–1249. 10.1007/s00170-010-2768-2, 2011.

[3] Richard P.A. Winter / Edgar Verscharen, Used is the better NEW @ Océ; Presentation held at Designprinzipien für die Kreislaufwirtschaft, Erkundungsreise mit Praxisbeispielen aus europäischen Unternehmen, Zürich, Donnerstag, June 28. 2018.

[4] IPCC, Changes in Atmospheric Constituents and in Radiative Forcing, 2018, https://www.ipcc.ch/publications_and_data/ar4/wg1/en/ch2s2-10-2.html, Retrieved April 10, 2020.

[1] European Commission, REGULATION (EC) No 1221/2009 OF THE EUROPEAN PARLIAMENT AND OF THE COUNCIL, on the voluntary participation by organisations in a Community eco-management and audit scheme (EMAS), repealing Regulation (EC) No 761/2001 and Commission Decisions 2001/681/EC and 2006/193/EC, 2009 (=EU Regulation (1221/2009/EC)), https://eur-lex.europa.eu/legal-content/en/ALL/?uri=CELEX%3A32009R1221, Retrieved April 10, 2020.

[2] FAO, Environmental Impact Assessment (EIA) and Environmental Auditing (EA), http://www.fao.org/3/v9933e/v9933e02.htm, Retrieved April 10, 2020.

[3] ClimateCalc, Tools for collection of data, https://www.climatecalc.eu/950.aspx, Retrieved April 10, 2020.

Chapter 9
Stakeholder and their contributions

As circular economy is cross-system and necessarily takes place across all stages of the value chain, it is necessary to develop, apply and ensure the adoption of company-wide concepts. In order to develop the right measures and ensure their implementation, the disciplines involved must be identified and their participation must be ensured. There is no one-size-fits-all solution, as depending on the region, industry segment and individual company, the addressed ways to achieve the goals may differ. In addition, sustainability strategies must be integrated and closely aligned with business strategies. Senior management has a role to play in driving the commitment to sustainability, providing support and monitoring the initiatives to achieve better results.

Different companies specialize into different specific capabilities, target groups and product portfolios. The knowledge needed in projects and product production is complementary – it is about aspects of the material and value flow as well as the profile of the product in the internal and external view. An interdisciplinary concept requires an interdisciplinary team with communication and collaboration to ensure knowledge exchange on all aspects of the product and its life cycle. As in all projects dealing with new tasks, the requirements for knowledge sharing and collaboration apply (openness, trust, shared vision, etc.). However, certain requirements need special attention, such as translation between disciplines and involvement of external partners in the supply chain or knowledge of and compliance with product regulations.

The interacting stakeholders in this context are as described in table 9.1.

Table 9.1: Interacting stakeholders in an organisation that produces material products.

- Strategy and planning/product design
- Sales and distribution
- Plant support
- Procurement and logistics/procurement partner management
- Logistics
- Manufacturing
- Research and development
- Quality assurance
- Legal
- Field service

To best achieve the goals, all stakeholders involved need to rethink product, design and functions. To do this best, all stakeholders need to work together with the same

https://doi.org/10.1515/9783110767308-009

vision – in some teams it helps to emphasise that the goal is not just an economic one but also one that is positive for the community and the environment.

The goals set are, at a high level, the same for all stakeholders and teams: minimising the environmental footprint, i.e. at least waste, energy consumption, material and toxin use and impact, water consumption, trying to allow the reuse of all materials and, enabling end-of-life management in a way that minimises environmental impact. This requires first identifying the key stakeholders, their objectives in this project and the high-level design activities to meet these objectives, as shown in the tables below.

It should generally be noted that collaborations with companies operating in the vicinity of the site under consideration can be very helpful in terms of joint purchase or harvesting of water and energy, mutual use of waste, exchange of ideas and more. Despite cross-company solutions, supporting local interactions often leads to effective and practical solutions that require limited organisational effort.

Identifying the stakeholders and their tasks and competences is crucial for the success of such a project. For this reason, some of the stakeholders are discussed in more detail below. It should be noted that the role of employees in reducing the environmental footprint should not be underestimated, as employee participation naturally influences organisational behaviour to reduce carbon emissions. Employees and their representatives also have the potential to contribute to the design of emission reduction opportunities in the workplace [1]. This source also lists a number of studies showing that the potential of employee knowledge and ideas can contribute to environmental management.

Analysis and implementation usually require a longer period of time, and it is often useful to work with supporting consultants. This is not necessarily because internal experts could not manage the task alone, but rather because of their experience and the internal pressure that an external consultant creates. The task to enlist the consultant for relates to preparation/team meetings with individuals/teams, enforcing timely delivery of results and updates, and monitoring and condensing what is done over time.

Questionnaires are presented below that address issues that the departments involved will discuss bilaterally or multilaterally. These lists are not meant to be exhaustive; they are intended to help individual teams get started in the discussions.

9.1 Manufacturing and logistics

In principle, manufacturing is simply the production of finished goods that the company sells to a customer after completion. Manufacturing uses raw materials or individual parts to obtain a product at the end. Obviously, the characters of manufacturing lines differ according to size, depth of production, standards considered mandatory, quality assurance measures, use of machinery and skilled labour, and many other parameters.

Quality control is crucial to manufacturing; quality refers to reproducible and predefined parameters that must be set in advance and adhered to from start to finish and relate to raw materials, production steps and the final product. This also refers to environmental regulations, whether internal or external. Therefore, quality control can play a crucial role in limiting the environmental footprint. Often, quality control is also the department that is primarily available to the auditors as an interlocutor during audits.

There are different types of production where different environmental aspects can come into play:

Make-To-Stock (MTS)

A factory produces goods to stock shops and showrooms. By anticipating the market for its goods, the manufacturer plans production activities in advance; the size of the mass increases the efficient use of energy and resources. The dilemma is that from an environmental footprint perspective this risks overproduction and implicitly wasting resources and energy (inbound and outbound logistics).

This dilemma is solved by the Make-To-Order (MTO) approach, where the producer stocks the production facilities, waits for orders before starting production when there is demand. This approach offers a reduced environmental footprint compared to the above, but often implies a longer waiting time at the customer's site and, depending on the type of product, frequent stopping of the production environment for changes in its set-up, often resulting in inefficient use of resources.

The Make-To-Assemble (MTA) approach, in which the factory produces components in anticipation of orders semi-finished products for later final assembly, does not solve this dilemma, or only to a limited extent. In principle, resources can be used somewhat more efficiently with this approach than with MTO – but if the orders fail to materialise, the manufacturer is left with an inventory of unwanted parts even with this approach.

Although none of these approaches fits every situation, it is useful to examine them as alternative courses of action to identify the optimal local approach.

As mentioned, the following is a collection of topics/questionnaires with content to be discussed bilaterally or multilaterally with other departments from a manufacturing perspective. Of course, this list should not be seen as exhaustive, but rather as a starting point and to clarify the direction of the discussion.

Restricted Substances Lists (RSLs) are usually provided to chemical companies to create an inventory of recommended maximum levels of known hazardous substances and restricted chemicals allowed on finished products and packaging in many industries. Examples include the RSL for the American Apparel and Footwear Association (AAFA), the Joint Industry Guide (JIG) for the electrical and electronics industries, and the Global Automotive Declarable Substances List (GADSL). The RSL should include a testing protocol with acceptable chemical limits and analytical methods to test products before they leave the factory.

The purpose of an RSL is to reduce the use of hazardous substances in consumer products and in the supply chain. Companies publish their RSLs to ensure that suppliers and vendors, as well as internal company employees, are aware of and can follow the restricted substance requirements. Also referred to as Green Procurement Standards, these are guidelines that suppliers must follow so that companies can better prepare for ever-evolving health and environmental standards by establishing and maintaining rules and documentation on the substances contained in their products.

Benefits of establishing a RSL (according to [2]):

- Compliance with various substance restriction laws such as RoHS, ELV, REACH, California Proposition 65, EU Packaging Directive, EU Batteries Directive, etc.
- Protect customer and consumer health and the environment
- Improving brand image by promoting health and environmental responsibility and sustainability
- Managing current and future chemical concerns throughout the supply chain and product life cycle
- Contributing to revenue protection by reducing the cost of recalls

9.2 Research and development

As indicated in the first chapters, in order to achieve sustainable production the consumption of material and energy, and the introduction of substances into nature must be reduced. In addition, the materials used and the energy required should come from renewable sources and be re-used as far as possible.

Therefore, R&D activities must focus on the eco-efficiency of products at all stages of their life cycle – i.e. to work with the aim of using less energy and less material applying design for re-use and recycling, and also to use what is consumed more efficiently, which subsequently leads to less waste. It is to be expected that costs borne by the general public (e.g. pollution and waste management) will sooner rather than later no longer be subsidised but borne by the polluters or "internalised" or passed on to industry and consumers [3]. For research and development, this means setting appropriate targets. As already mentioned, the minimum expectation is that not only is a commercially viable product proposal required, but that the same proposal is also ecologically viable. The latter means that the product is not only characterised by its price in terms of money, but also by its price in terms of environmental footprint. Both prices are parameters for the commercialisation and marketing of the produced supply.

Consequently, research and development objectives must vary to include appropriate outcomes at a competitive level for all stages of the product life cycle. As these types of value tags are not available for most of the competitive landscape, it is reasonable to assume that the development of appropriate value schemes and discussion of them will become a key activity of strategy and R&D in the coming years.

The following Table 9.2 is a non-exhaustive collection of topics/questionnaires with content to be discussed bilaterally or multilaterally with other departments from an R&D perspective.

Table 9.2: High-level cooperation topics for manufacturing.

Target	Analysis and design of	Jointly with
Minimising waste	– Allow for repair, maintenance, recycling or take-back service programmes	Research and Development, Field Service,
	– Document disassembly in comprehensible manuals (assume that disassembly is carried out elsewhere by less qualified personnel abroad with whom it is not possible to communicate)	Manufacturing, Marketing
	– Develop an integral parts and materials strategy to enable parts to be reused and refurbished or taken back as raw material	
	– Implement a materials strategy that avoids composites and alloy variants to increase the value and recycling rate of product parts	
	– Set up production facilities and infrastructure so that material and energy consumption can be measured	
	– Enable the return of used goods (and parts) after use to reduce material consumption and costs at this stage of the life cycle	Manufacturing, Field Service, Facility Management
	– Use renewable energy in production, testing, refurbishment, internal transport and use	Procurement, Research and Development
	– Use recycled and recyclable materials for purchased consumables/spare parts	Manufacturing, Field Service, Research and Development
	– Opt for removable technical coatings to reduce wear and tear in the use phase	

(continued)

Table 9.2 (continued)

Target	Analysis and design of	Jointly with
Minimising consumption of energy	– Use of renewable energy sources, especially for electricity, heating and cooling	Facility Management
	– Targeted production planning to schedule unproductive times (e.g. service times – leading to production shutdown) to times when renewable energy is not available	Facility Management, Manufacturing
	– Consider measures to reduce energy consumption for consumers/users	All
	– Ensure simplification of disassembly	Research and development, field service
	– Allow a second life (or more) after refurbishment.	
Reduction of throughput of chemical materials	– Select appropriate materials that use fewer chemicals from the raw material stage to production or release fewer or no substances of concern during their lifetime	Procurement, Research and Development, Quality Management
	– Development and adoption of a Restricted Substance List (RSL)* that reflects the strategy to reduce these substances and also considers the health impacts on workers and consumers	Procurement, Research and Development, Health and Safety, Quality Management
	– Use of mechanical as opposed to chemical technologies	Research and Development, Field Service

End-of-life management	– Plan for storage, record keeping, provision and testing of reusable/ remanufactured parts/materials	Facility Management
	– Plan for appropriate storage of returned materials to be sold/reused; prepare infrastructure for disassembly	
	– Select appropriate materials that can be recycled/remanufactured or reused	Facility Management, Health & Safety, Quality Management, Procurement
	– Avoiding the use of mixtures and composites that cannot be separated	Research and Development
	– Incorporate features that can be repaired, replaced or upgraded to maximise durability and longevity	Research and Development, Procurement
	– Preferably use materials that are produced in closed-loop systems	Strategy, Research and Development, Marketing,
	– Do not hinder recycling/industrial biodegradability	Procurement, Research and Development
Reduction of the consumption of hazardous substances and water	– Enabling risk management in cycles by providing information on substances of concern along the value chain(s) (such as CMR (carcinogenic, mutagenic or toxic for reproduction) or PBT (persistent, bio-accumulative and toxic substances)	Manufacturing, Procurement, Research and Development, Marketing, Health and safety
Misc.		

*A Restricted Substance List (RSL) and Declarable Substances List lists substances or chemicals that are restricted, for example, by government regulations or laws. Some companies define internal standards or guidelines that also define an internal RSL. This is useful, for example, if a company operates internationally and needs to comply with different regional regulations and anticipate upcoming changes so that they go beyond local legal requirements. Often brands or trade associations have their own RSL. Managing and updating an RSL is difficult as it requires constant monitoring of the laws.

9.3 Facility management

Building management is responsible for ensuring that the systems of the built environment or facility function harmoniously and efficiently. It ensures that the places where people work, play, learn and live are safe, comfortable, productive and sustainable. To this end, it maintains land, buildings, facilities and other environments that house staff, productivity, inventory and other elements of operations. Basic tasks include mitigating a company's environmental impact and promoting sustainable tactics for long-term cost management – also to ensure compliance with regulations [4]. In this context, facility management organises to shift supply towards lower consumption of energy and resources and a reduction of waste (in terms of toxicity and mass, but also in terms of reusability). Procurement is one part of this – e.g. switching energy supply to as much renewable energy as possible, e.g.

1. Buying only 100% renewably generated electricity;
2. Covering heating by geothermal energy, i.e. using hot water from a depth of often several 1,000 m for heating;
3. Cooling via a central cooling system at the site or jointly with neighbouring companies, e.g. by using groundwater – which can lead to a site (or neighbouring companies) needing up to almost no electrical cooling systems.

Table 9.3 is a non-exhaustive collection of topics/questionnaires with content to be discussed bilaterally or multilaterally with other departments from a facility management perspective.

9.4 Procurement

Procurement is the process of seeking and agreeing terms and conditions and acquiring goods, services or works from external sources, often through a tender or competitive bidding process. To analyse the procedures used and outcomes in procurement, it is good practice to use economic analysis methods such as cost-benefit analysis when sound data is available [5]. Climate and green procurement that takes into account the carbon footprint of products and services in e.g. Green Public Procurement (GPP) can help reduce consumption and waste. As seen above, collecting, assessing and comparing data is difficult. As a result, the technical complexity of Green Procurement (GP) combined with low administrative capacity, the use of supplier audits and the building of specialised procurement capacity is a barrier to implementing green procurement.

Just as the company for which procurement operates requires environmental footprints for all products manufactured, procurement must insist on a statement of an energy and material footprint for all products purchased, including emissions and waste.

Table 9.3: High-level cooperation topics for research and development.

Target	Analysis and design of	Jointly with
Minimising waste	Development of	
	– Approaches and systems that enable repair, maintenance, recycling or take-back programmes	Field Service, Manufacturing, Marketing, Strategy
	– Material strategies that avoid composite materials and alloy variants to increase the value and recycling rate of product parts	Procurement, Manufacturing,
	– Design assumptions that enable and monitor the material and waste streams generated during the product life cycle	Marketing
	– For product-market combinations where this is possible: develop in a way that encourages use in conjunction with biodegradable/recyclable materials	Marketing, Field Service
	– Reduce the variety of materials used	Marketing, Strategy
	– Improve long-term usability	Manufacturing
	– Enable sharing of a product	Marketing
	– Integrate multiple functions into a single product (such as a smartphone)	Marketing, Manufacturing
	– Enable repair, maintenance and recycling programmes	Manufacturing, Field Service, Marketing
	– Enable an integral parts and materials strategy to allow parts to be reused and refurbished or returned as raw materials	Manufacturing, Field Service, Marketing
	– Implement a materials strategy that avoids composites and alloy variants to increase the value and recycling rate of product parts	Manufacturing
	– Set up production facilities and infrastructure so that material and energy consumption can be measured	Facility Management, Manufacturing Marketing, Filed Service,
	– Enable the return of used goods (and parts) after use to reduce material consumption and costs at this stage of the life cycle	Manufacturing, Procurement,
	– Selecting removable technical coatings to reduce wear and tear in the use phase	Facility Management, Manufacturing Manufacturing, Field Service
	– Selecting appropriate materials (including organic and recycled and recyclable)	Field Service, Manufacturing

(continued)

Table 9.3 (continued)

Target	Analysis and design of	Jointly with
Minimising energy consumption	– Use renewable energy during production, testing, refurbishment, internal transport and use.	Manufacturing
	– Enable a second life (or more) after refurbishment.	Manufacturing, Field Service
	– As applicable: enable sequential start-up of sub-units to reduce start-up current.	Manufacturing
Reduced use of chemical substances	– Selecting appropriate materials that use fewer chemicals from the raw material stage to production or release fewer or no substances of concern during their lifetime.	Manufacturing, Procurement
	– Development and adoption of a Restricted Substance List (RSL)* that also takes into account the health impacts on workers and consumers	Manufacturing, Procurement, Health and Safety, Legal, Quality Management,
	– Use of mechanical as opposed to chemical technologies	Manufacturing, Field Service
	– Removal of toxic substances (compliance as a minimum, in line with requirements such as RoHS and REACH directives)	Manufacturing
End-of-life management	– Preparation of the infrastructure for dismantling	Manufacturing, Quality Management, Facility Management
	– Selecting suitable materials that can be recycled/recycled or reused	Manufacturing
	– Avoid using mixtures and composites that cannot be separated	Manufacturing, Procurement
	– Preferably use materials that are produced in closed loops	Strategy, Manufacturing, Marketing,
	– Do not hinder recycling/industrial biodegradability	Procurement, Manufacturing

Reduced consumption of hazardous materials and water	– Developing a modified materials strategy (to reduce the number of materials).	Procurement, Manufacturing, Field Service
	– Use of mechanical technologies as opposed to chemical ones	Manufacturing
	– Selecting appropriate materials (including organic and recycled and recyclable) to allow for optimal life cycle use	Manufacturing, Procurement
	– Developing a materials strategy that enables multiple use and reuse of materials	Manufacturing, Procurement
	– Choice of engineered coatings to reduce wear and tear during the use phase	Manufacturing,
Minimising energy consumption	– Consider ways to achieve zero energy for office environments, laboratories, R&D and production	Procurement, Facility Management
	– Consider collaboration with nearby factories and ways to work with them to meet guidelines	Procurement, Facility Management
	– Reduce number of parts and tools	Manufacturing
	– Focus on long life and ease of maintenance	

Table 9.4: High-level cooperation topics for facility management.

Target	Analysis and design of	Jointly with
Minimise waste	– Establish production facilities and infrastructure to measure the consumption of materials and energy	Facility Management, Research and Development
	– Use of renewable energy in production, testing, refurbishment, internal transport and use of equipment and infrastructure	Procurement, Facility Management, Research and Development
Minimising consumption of energy	– Set up infrastructure so that consumption of materials and energy can be measured	All – Facility Management in Lead
	– Monitor material and waste flows; minimise waste, maximise reuse.	
	– Manage energy and waste	All – Facility Management in Lead
	– Account for flows	
	– Automatic inventory update to mobilise collection, tracking of recycled materials	All – Facility Management in lead
	– Set up and optimise infrastructure for production, end-of-life management/handling, administration, research and development, etc.	
	– Prepare infrastructure for collection of parts, systems and waste from involved parties and for recycling. Collection of returned systems and parts	Distribution and Logistics
	– Organise and monitor waste management	Manufacturing
	– Minimising energy consumption. Aiming for a zero energy environment for production	Procurement
	– Carbon footprint-free supply of energy (el. energy: wind, solar panels, external supply with certified sources) and heating/cooling: depending on local possibilities (groundwater, deep water (heating)) to reduce the consumption of materials and water	Facility Management, Procurement, Research and Development
	– Enabling the return of used goods (and parts) after use to reduce material consumption and costs at this stage of the life cycle	Procurement, Facility Management, Research and Development
	– Using recycled and recyclable materials when procurement consumables/ consumables	Facility Management, Research and Development
	– Setting up production facilities and infrastructure to measure the consumption of materials and energy	Procurement
	– Use renewable energy in production, testing, refurbishment, internal transport and use	Facility Management, Research and Development

Minimise consumption of energy (see also above)	– Striving for a zero-energy environment for administration, research and development and production – The use of renewable energy sources – especially for electricity, heating and cooling – Implementing special production planning to schedule non-productive times (e.g. service times – leading to production shutdown) to times when renewable energy is not available	Procurement, Facility Management Facility Management
Reduced use of chemistry	– Use recycled and recyclable materials for your consumables/spare parts	Manufacturing, research and development
End-of-life management	– Prepare infrastructure for disassembly – Plan for storage, record keeping, provision and testing of reusable/remanufactured parts/materials – Plan for appropriate storage of returned materials and materials to be sold/reused	Manufacturing, Quality Management Manufacturing, Facility Management, Manufacturing, Health & Safety, Quality Management, Procurement

Materials, energy and products with a low environmental footprint, if the labelling is available and tangible, are often perceived as more valuable on the customer side – therefore establishing GP without burdening the operating and net profit is an important metric for procurement. Although it is an option to consider GP as an additional burden on the balance sheet, it should be seen as a competitive advantage. Managing suppliers is a key measure in implementing a sustainable procurement programme. According to [6] it is beneficial to

- develop characterising standards to be applied to partners to ensure their performance on sustainability, often e.g. in "sustainability meetings" with suppliers to promote and improve sustainability metrics.
- Establish a "Supplier Code of Conduct" with sustainability metrics against which suppliers can be objectively audited (see Quality Management).

There are many assessment tools and techniques used by procurement teams to evaluate supplier performance on sustainability. For example, apparel and retail companies use the Higg Index to assess their suppliers' performance against sustainability targets and also work with third-party risk management companies such as Sedex, EcoVadis, Ecodesk and Cr360. Essentially, these indices obtain ratings for organisations and compare them to the good practices followed to improve supplier sustainability practices. In the context of managing suppliers, contract management plays a crucial role in enabling procurement professionals to get suppliers to undergo audits and behave in a compliant manner. Procurement organisations need to include requirements related to supplier audits when contracts are renewed. This helps procurement management to identify only those suppliers who are willing to submit the results of appropriate audits when required.

Table 9.4 is a non-exhaustive collection of topics/questionnaires with content to be discussed bilaterally or multilaterally with other departments from a procurement perspective.

9.5 Marketing

In this context, marketing encompasses the processes involved in the interaction between product and customer. It also includes the identification, selection and development of a product, part of the product design, its price, the channel through which it reaches the customer, and the development and implementation of a promotional strategy. As indicated above, about 80% of a product's environmental impact is determined during its design phase. Both the identification/development of the product and the communication about it may change if sustainability-related issues become more important in the eyes of customers or if prices increase due to rising material and/or energy consumption. Environmental concerns, such as those about global warming and the shortage of raw materials, increase the demand effect and thus

Table 9.5: High-level cooperation topics for procurement.

Target	Analysis and design of	Jointly with
Minimise waste	– Develop a materials strategy that avoids composites and alloy variants to increase the value and recycling rate of product parts – Use of renewable energies in production, testing, remanufacturing, internal transport and use – Use renewable energy in production, testing, remanufacturing, internal transport and use – Use recycled and recyclable materials for the consumables you purchase – Procurement of sustainable materials – Establish an ERP system that is able to differentiate between product and non-product related materials, energy consumption and waste generation – Focus on fewer materials and "made to last" – Support internal and partners to offer programmes for repair, maintenance, recycling or take-back services – Challenge partners to zero-waste design/pattern design that increases efficiency and leads to cost savings (more revenue per . . .) – Develop a materials strategy that avoids composites and variations in alloys to increase the value and recycling rate of product parts – Use renewable energy in production, testing, remanufacturing, internal transport and use	Manufacturing, Marketing Facility Management, Research and Development Research and Development, Procurement Procurement, Research and Development, Strategy, Quality Management Quality Management Quality Management Procurement, Manufacturing, Marketing Procurement, Facility Management, Research and Development
Minimising consumption of energy	Striving for a zero-energy environment for administration, R&D and production. Selecting appropriate materials (including organic and recycled and recyclable) that allow for optimal use in production Selection of technical coatings to reduce wear and tear during the use phase Considering the strength of environmental regulations and infrastructure at the site of suppliers and factories, and how to work with them to meet the guidelines	

(continued)

Table 9.5 (continued)

Target	Analysis and design of	Jointly with
To reduce chemical input and impacts	– Selection of appropriate materials that utilise fewer chemicals from raw materials phase to production or release lesser or no substances of concern during lifetime – Develop and adopt a Restricted Substance List (RSL)* that also considers health impacts to workers and consumers Adopt policies for Restricted Substance List (RSL) that considers health impacts to workers and consumers while working with suppliers/factories to support compliance Use of mechanical technologies as opposed to chemical	Manufacturing, R&D Manufacturing, R&D, Health and Safety, Quality Management, Legal All All
For end-of-life management	– Plan for fitting storage of returned and to be sold/reused materials – Avoid the use of blends and compound materials, that cannot be separated – Preferably use materials that are produced with closed-loop systems – Not hinder recycling/industrial bio degradability	Facility Management, Health & Safety, Quality Management, Manufacturing R&D, Manufacturing Manufacturing, R&D Manufacturing, R&D
To reduce consumption of hazardous wastes and water Misc.	Monitor installed base for reused parts Buy renewable supplies electrical energy Try to cooperate with neighbouring companies on joint harvesting of energy resources (heat, fresh water) Review opportunity for the use of ground water for cooling and geothermal energy	Field Service Procurement Procurement Procurement

create an incentive to engage in sustainability in a cost-increasing way. At the same time, the cost effect motivates to increase the price, which can lead to lower demand. Taken together, these demand-side and supply-side effects jointly determine the relative profitability of adjusting environmental footprint and price.

When the demand effect outweighs the cost effect, increased environmental awareness can be an incentive to develop a greener product with a lower environmental footprint. Daniel Halbheer [7] nicely summarises how marketing can capitalise on environmental concerns and the attention they generate: Based on the observation that technology, consumer demand, corporate willingness and government legislation are likely to make footprint labelling a reality in the near future, Halbheer places the label at the centre of his consideration, since in his view labels provide a way to communicate about environmental concerns. As mentioned earlier, in addition to defining the product, marketing is also about the value propositions associated with products that are received as "green". Based on the review of the literature and our own analysis, several value propositions that are coming to market or are present have been identified, not all of which apply to all products, but can serve as a guideline for what marketing can add in product definition and communication – important ones are listed in Table 9.6 below.

9.6 Quality management

Quality management and control (QM) is critical to the manufacture and performance of the product; quality refers to reproducible and predefined parameters that must be established upfront and maintained from the beginning to the end of the product's life and relate to raw materials, production steps and the final product. This also relates to environmental regulations, whether internal or external. Therefore, quality control can play a crucial role in limiting the environmental footprint. QM is applied to incoming and outgoing products and processes, and in some companies also to external partners and suppliers, distribution and logistics. In this context, QM has the task of reviewing and auditing what has been negotiated by Procurement with partners. QM develops and implements expertise and checks required for the audits to be carried out. These can include sustainability scorecards, supplier evaluation measures for sustainability parameters and tracking methods for a supplier's sustainability performance, conducting sustainability meetings to promote and improve sustainability parameters, and establishing a supplier code of conduct against which suppliers can be objectively audited.

The following Table 9.8 provides a non exhaustive collection of topics/questionnaires with content to be discussed bilaterally or multilaterally with other departments from the QM perspective.

Table 9.6: Additional propositions of material products to be considered after reduction of environmental footprint.

Expected Propositions of Products	
1.	Consumers will have more trust in a governmentally regulated environmental footprint label than a voluntary industry-led one.
2.	In comparing two similar products, consumers will weight environmental footprint over emissions alone.
3.	In comparing two similar products with the same Eco Footprint, consumers will prefer lower environmental footprint over higher Offsets.
4.	When comparing two similar products with the same Offsets, consumers will put more weight on Recycling than on Carbon Sinks.
5.	When comparing two similar products with the same Offsets, consumers will prefer products from firms that manage their own Offset programs to those from firms that outsource their Offset programs.
Expected Propositions of Companies	
1.	Firm technological innovation and environmental footprint level will interact to affect firm performance; high innovation firms with low environmental footprint will have higher financial performance and less financial risk than others.
2.	The extent to which firms lower their environmental footprint will be moderated by the extent to which consumer reactions affect the firm. Firms that are less affected by consumer reaction will be less likely to lower their Eco Footprint.
3.	Firms that are more distant from customers (e.g., B2B firms) will expend fewer resources toward reducing carbon emissions than will firms that are directly involved with consumers (e.g., B2C firms).
4.	As a firm's level of market orientation increases, its product's environmental footprint will decrease.
Expected Propositions of Target Groups	
1.	Females will use environmental footprint labels to make product decisions more than males.
2.	More highly educated consumers will use environmental footprint labels to make product decisions more than less educated consumers
3.	Younger consumers will use environmental footprint labels to make product decisions more than older consumers.
4.	The attention consumers pay to environmental footprint across the four stages of the product life cycle will be moderated by locus of control: Consumers with higher levels of internal locus of control will pay more attention to the environmental footprint label than those with lower levels of internal locus of control. Consumers with higher levels of internal locus of control will place higher emphasis on the recycling part of the environmental footprint label compared to those with lower levels of internal locus of control.

Table 9.6 (continued)

5.	Consumers with higher levels of symbolic moral identity will place greater emphasis on environmental footprint labels when making product decisions than those with lower levels of symbolic moral identity.
6.	Consumers' prior expectations of the producer's environmental footprint will moderate the effect of the perception of a product's environmental footprint level on their purchase decision; a product with the same level of environmental footprint will be preferred more if the customer has low expectations than if the customer has high expectations.
7.	The greater the amount of attention one's peers pay to carbon emissions the more the individual will pay attention to Eco Footprint.
8.	Collectivist cultures will pay more attention to environmental footprint than individualist cultures.
9.	Group adoption of environmental footprint labels will moderate belief in the efficiency of environmental footprint sinks. As group adoption of environmental footprint labels increases, so will the emphasis placed on environmental footprint sinks.
10.	Cultural style will moderate the relationship between peer influence and attention paid to environmental footprint labels. As societal collectivism increases the influence of a peer group will decrease.
11.	The economic level of a group will moderate the relationship between peer influence and attention paid to environmental footprint labels. As the level of economic disadvantage increases the influence of a peer group will decrease.
12.	The usage part of the environmental footprint label will carry more weight for utilitarian than hedonic products.
13.	As purchase frequency increases the environmental footprint label will play less of a role in product choice.
14.	Carbon footprint improvements beyond the product category average will produce positive but diminishing marginal financial returns.
15.	Increases of environmental footprint levels above the product category average will produce asymmetric (accelerated) negative financial effects

Note that the restriction made in [7] to "carbon emissions" is considered inapplicable, as in this context environmental footprint and waste must be used as the identifying label:

The following Table 9.7 provides a non-exhaustive collection of topics/questionnaires with content to be discussed bilaterally or multilaterally with other departments from a procurement perspective.

Marketing can and should inform the customer about the environmental design principles set by the manufacturer – i.e. what has been focused on in the product design phase; topics and premises in this communication are critical for success. The following table provides an overview of possible and noteworthy goals and strategic approaches, e.g.:

1. Reduction of energy consumption during supply chain, manufacturing and use.
2. Reduction of material consumption during supply chain and manufacture e.g. the possibility to use the system's ability to use recycled and recyclable material or industrially compostable materials.
3. The development of a parts and materials strategy to enable reuse and remanufacturing, and the application of specific design measures and services to extend the lifetime (of the system itself and of the parts) and to enable a second life (or more) after refurbishment.
4. A materials strategy that avoids composites and variations of alloys to increase the value and recycling rate of the product parts.
5. Implement systems that allow remote access to machines to enable predictive maintenance to enable predictive maintenance.
6. Design replacement parts in such a way that they can, in principle, be remanufactured and tested on a large scale in order to and tested on a large scale to enable multiple use. Accordingly, the service concept should allow system modules or the entire system to be reconditioned either on site at the customer's site or in the factory with the aim of enabling a second product life.
7. Implement a materials strategy that avoids composite materials and a wide variation of alloys in order to increase the value and recycling rate of the product parts while still being able to do the job in design and operation. The material selection allows for high-return rates, limiting the amount of scrapped material and landfill are limited.

9.7 Health and safety

The Environment, Health and Safety (EHS) department focuses on identifying and recording all laws, rules, guidelines and processes that help protect employees, the public and the environment from harm. The topic area includes for the design and implementation of appropriate procedures at workplaces for environmental protection, occupational health and safety, such as the preparation of safety data sheets, product safety data sheets or declarations of conformity. From an occupational health and safety point of view, EHS management means, among other things, creating procedures to identify workplace hazards and reduce accidents and exposure to harmful situations and substances.

From an environmental point of view, it is about creating a systematic approach to environmental compliance. Sub-targets are managing waste or air emissions, reduce the company's environmental footprint and defining an environmental, health and safety (EHS) policy to limit environmental impacts, improve occupational safety, meet internal standards such as conducting an annual environmental risk assessment at operational sites or conducting additional audits at complex or high-risk sites. Simplified targets for health and safety are sketched in Table 9.9 below

Table 9.7: High-level cooperation topics for marketing.

Target	High-level design activity	Together with .
Misc.	– Enable repair, maintenance, recycling or take-back service programmes	R&D, Field Service, Marketing
	– Document disassembly in understandable manuals (start with the assumption that disassembly is performed somewhere else by lesser qualified personnel abroad which cannot be communicated with)	Marketing
	– Enable to take back of used goods (and parts) after use to reduce material consumption and cost in that phase of life – Think of advanced and fierce competitors	Marketing, Filed Service, R&D,
	– What is communicated with respect to sustainability, and how competition – Defines – Implements – Practically does – Measures their care about Umweltfußabdruck	Legal, Strategy
	– Identify target group of communication (own/competition) – Which strategies are followed, which values are used, is there something to learn from?	Legal, Strategy, R&D
	– Define benefits to participants in value chain, and, of course, the environment. – Address product impacts and benefits across the lifecycle	Sales
	– Develop own – and true – propositions for the company and the product plus communication strategies around them – Communication about – design for disassembly, recyclability, reuse, remanufacture ("cradle to cradle" thinking)	Sales, Strategy
	– Research legal development with respect to – Enforced tracking of materials – Taxation for/enforces return of wasted materials – Handling of returned materials in market/regions	Sales, Strategy

Table 9.10 provides a non-exhaustive collection of topics/questionnaires with content to be discussed bilaterally or multilaterally with other teams from the perspective of the Health and Safety Department.

9.8 Legal

Legislation and case law change as climate change progresses or to ensure that climate change is minimised. Therefore, there is often a need to reassess or ensure that business activities are compliant with the new or anticipated legal situation [8] or that contracts with suppliers reflect the changed situation.

Some examples where this may be relevant:
- As a rule, the purchase of new material is agreed in contracts – as described above, in order to reduce emissions, it may make sense to use used material instead of new raw material for the entry into future value creation cycles. This situation must be reflected in contracts.
- In order to get access to used machines and components, it may make sense not to sell the machine or the components to the customer, but to conclude an agreement of use for an unlimited period of time, whereby the machine is returned after use.
- As described above, companies will probably be forced to achieve a defined reduction of their emissions. In many value chains this will not be possible without pressure to reduce emissions from processes and raw materials in the supply chain. This must be agreed with suppliers
- emissions or material consumption must be reduced – the question of who assumes liability if, for example, delivery failures occur due to non-compliance with the relevant contracts or regulations must be clarified.
- Supply chains are largely international. The protection according to different legal norms must be legally secured.

Table 9.8: High-level design activities for quality management.

Target	High level design activity	Together with
To minimise waste	– Ensure consistent quality – Despite reduction in materials used and material diversity – When using refurbished parts, recycled materials – Externally: support Procurement in partner audits	Manufacturing, R&D Procurement
To reduce chemical input and impacts	– Develop and adopt a Restricted Substance List (RSL)* that also considers health impacts to workers and consumers	Manufacturing, Procurement, Health and Safety, Quality Management, R&D, Legal

Table 9.11 displays a non-exhaustive collection of topics/questionnaires with content to be discussed bilaterally or multilaterally with other departments from the perspective of the lawyers.

9.9 Distribution and logistics

Distribution and logistics have the task of receiving, storing or shipping ordered items and routing them efficiently. Generally, there is inbound and outbound distribution;

inbound checks that the goods received match the order and are in the right quantity and quality and routes the goods accordingly. From there, the goods are taken to a warehouse and the paperwork is forwarded. The same logic applies to outbound logistics – only in reverse. In connection with access to the raw materials sold, a distribution and logistics department must focus on returns and the corresponding infrastructure in addition to the tasks pursued so far. The return policy is complex because

- the customer must agree in advance to the return of the system
- the user/installation characteristics must be taken into account
- a collection and sorting process must be in place to distinguish parts that are to be reused/unmodified, tested and repaired, refurbished or recycled to serve as a primary resource [9]
- the internal and external members of the supply chain, such as customers, suppliers and logistics centres, must be taken into account

In the context of the environmental debate, the term Green Logistics was created. It describes alternative ways for logistics to fulfil its tasks in an environmentally friendly and sustainable way. The focus is on the use of environmentally friendly means of transport, route and storage planning that is also cost-optimised in terms of emissions, and the use of cooperation with partner companies (e.g. ownership of different means of transport or storage locations) in the supply chain – the calculation of the emissions caused by transport (often including third-party facilities and means of transport) on the shipping quantities of individual parts or product lines may be necessary for this purpose. Often Green Logistics is computerised, e.g. using simulations to optimise storage locations or assess the carbon footprint across the supply chain.

Distribution and logistics also need to plan for supply chain failures – this will become increasingly important as the expected global productivity losses from the effects of climate change will be around one third of potential performance per year, according to a warning from the UN Secretariat [11].

Energy costs account for a significant part of distribution and logistics costs. One of the goals of Green Logistics is to assess, measure and reduce the emissions generated by storage (location of goods, means of access to them) and transport (electricity consumption). The decision to produce continuously or in batches also strongly influences the storage and transport concepts used.

Table 9.12 provides a non-exhaustive collection of topics/questionnaires with content to be discussed bilaterally or multilaterally with other departments from the perspective of the Distribution and Logistics team.

Table 9.9: Targets for health and safety.

Target	Such as	Normalised by
Achieve third-party standards	ISO 14001	n.a.
Certification at x% of global manufacturing facilities		
Energy consumption	x*% change	Annual production (units) per employee per product (type a, b, c)
Water usage	x% reduction	Annual production (units) per employee per product (type a, b, c)
Hazardous waste disposal	x% reduction	Annual production (units) per product (type a, b, c)
Solid waste disposal:	x% reduction	Annual production (units) per product (type a, b, c)
Waste diversion rate in – owned operations – partner operations	x% reduction	Annual production (units) per employee per product (type a, b, c)
Greenhouse gas emissions	x*% change	Annual production (units) per product (type a, b, c)

*assume a target of 0 by 2040 latest

9.10 Field service

Companies offer field services to maintain systems that are operated at the customer's site. This type of service, when properly optimised, allows for the extension of the life of products. In addition, the available digital infrastructure may be able to provide statistics on the expected life of parts and systems. This data allows the exchange (and supply of refurbished parts), maintenance and facility service to be planned in an emission-saving way to prepare the infrastructure for refurbishment. Fed with sensory information from the system, the first fix rate increases, scheduling of appointments minimises distances and fuel consumption while reducing the need for repeat visits. Using the existing infrastructure, optimised routes for logistics can be calculated to save avoidable return trips and unnecessary detours. This helps to save time and fuel.

Table 9.13 provides a non-exhaustive collection of topics/questionnaires with content to be discussed bilaterally or multilaterally with other departments from a Field Services perspective.

Table 9.10: High-level cooperation topics for health and safety.

Target	High-level design activity	Together with
To minimise waste	– Enable repair, maintenance, recycling or take-back service programmes	R&D, Field Service, Marketing
To minimise waste	– Remove toxic substances (compliance as a minimum, in line with requirements of RoHS and REACH directives)	
To reduce chemical input and impacts	– Develop and adopt a Restricted Substance List (RSL)* that also considers health impacts to workers and consumers – Plan for reducing environmental impacts from the outset	Manufacturing, Procurement, Health and Safety, Quality Management, Legal

9.11 Duties of all stakeholders

As indicated, employee participation has been found to influence organisational behaviour to reduce carbon emissions [12]. Representatives also have the potential to contribute to the design of workplace emission reduction opportunities. The following Table 9.14 displays some high-level design activities involving all stakeholders.

Table 9.11: High-level cooperation activities for legal.

Target	High-level design activity	Together with
	– Ensure that the contracts with partners up and downstream the supply chain permit take-back service programmes to the advantage of the company	All
		All – Legal in lead
	Contractual arrangements for liability and re-access of raw materials – Ensure ownership as far as reasonable, and – Ensure the right to access used materials to ensure access in larger quantities	All – Legal in lead
	– Research and communicate risks due to changed legislation and litigation – Ensure that wording of offers are phrased so that they fit to latest status of and expectations in legislation and litigation when containing information on the product's	All – Legal in lead

Table 9.11 (continued)

Target	High-level design activity	Together with
	– Environmental footprint and – Legal conditions around that (liabilities, e.g. for environmental footprint and especially emissions throughout lifetime of product, tasks of customer and vendor, ownership of the material contained)	
To reduce chemical input and impacts	– Develop and adopt a Restricted Substance List (RSL)* that also considers health impacts to workers and consumers	Manufacturing, Procurement, Health and Safety, Quality Management,

Table 9.12: High-level cooperation topics for distribution and logistics.

Target	High-level design activity	Together with
	– Prepare infrastructure for collection of parts, system, waste from parties involved and recycled. Collection of returned systems and parts – Organise the area for waste after disassembly and monitor it – Use environmentally friendly means to transport and store – Optimised planning of routes and means for transport and storage – For that also cooperation with partners – Map and monitor emissions per product/portfolio, region – Use simulations to optimise locations of storage or assess carbon footprints throughout the supply chain.	Facility Management
Packaging	– Optimisation of packaging (e.g. packaging to product ratio) – Link to targets on waste minimisation (e.g. reduction of packaging weight per kg of product) and link to industry initiatives/context to set targets – Use of "low impact" packaging materials (e.g. recycled/ recyclable/renewable), which also have a lower environmental footprint – Design packaging for recycling/reuse – Packaging system protects product from damage to extend life – Packaging includes information allow the end user to get the maximum value and utility	

Table 9.12 (continued)

Target	High-level design activity	Together with
Transportation	– Products are transported using the most economical forms of transport available with low environmental impact – Use of "low impact" transit packaging (e.g. recycled/recyclable/renewable), which also have a lower environmental footprint – Optimisation of transit packaging without exposing product to damage	

Table 9.13: High-level cooperation topics for field service.

Target	High-level design activity	Together with
To minimise waste	– Enable repair, maintenance, recycling or take-back service programmes – Monitor consumption and waste during lifetime – Enable integral parts and material strategy to permit reuse and refurbishment of parts or returns as raw material – Enable to take back of used goods (and parts) after use to reduce material consumption and cost in that phase of life – Chose for removable technical coatings to reduce wear and tear during use phase	R&D, Field Service, Marketing Manufacturing, Marketing R&D, Field Service, Marketing, Filed Service, R&D, R&D, Field Service
To reduce consumption of energy	– Test measures to reduce consumption of energy for consumers/users – Ensure simplification of disassembly – Enable a second life (or more) after refurbishment.	R&D, Field Service
For end-of-life management	– Define and execute service cycles fitting to age of product (as required)	

Table 9.14: High level cooperation topics for all stakeholders.

Target	High-level design activity	Together with
	– Start bookkeeping and allocating consumption and waste to individual projects/ products/product lines – Set up dedicated ERP system to enable product related monitoring of materials related to production and produce throughout phases – Select of appropriate materials that utilise fewer chemicals from raw materials phase to apparel production or release substances of concern during incineration/in landfills	All – Procurement, R&D and Manufacturing in lead
	– Adopt policies for Restricted Substance List (RSL) that considers health impacts to workers and consumers while working with suppliers/factories to support compliance	All – R&D and Field Services in lead
	– Participate in predictive LCA of products-to-come. Reduction of environmental footprint is the main focus, but also keeping value in the company by implementing a coherent material strategy.	All – R&D and Field Services in lead
	Analysis of design options to reduction of environmental impact of product when in production, use, or reuse: – Ensure design for disassembly, recyclability, reuse, remanufacture (avoid “cradle to grave”- thinking) – Prevention of use of harmful, toxic, forbidden materials/ of unclear or prohibited origin in all phases of life – Develop and implement means to extend lifetime of the good produced (of the system itself and the parts) – Enable a second life (or more) after refurbishment – Reduction of the use of water internally and users and consumers	

To minimise waste	– Set up infrastructure so that consumption of materials and energy can be measured – maximise reuse – Manage energy and waste – book keeping of flows – Automatic inventory updates to mobilise pick-up, tracking of recycled materials – Set up and optimise infrastructure for production, end-of-life management/handling, administration, R&D, etc.	All – Facility Management in Lead All – Facility Management in Lead All – Facility Management in Lead
To increase ethical production	– Incorporation of traditional, local and global craft skills that advocate for worker rights – Collaboration with organisations that promote environmental stewardship and worker rights – Implement policies or programmes including audits for suppliers and factories to increase transparency and accountability within the supply chain	All – Manufacturing in leadAll – Strategy in lead All – Procurement in lead
End of life management	– Set up processes and services fit for product take-back and recovery	All – Manufacturing in lead
To reduce consumption of hazardous wastes and water	– Identify and quantify substances of concern in products, materials and components – Reduce/substitute substances of concern – give priority to substances with a better profile regarding lower toxicity, lower persistency, etc. and substitute CMR*/PBT with non-CMR/non-PBT substances	All – Manufacturing, R&D in lead

References

[1] Reducing carbon emissions through employee participation: Evidence from Australia; Markey, R., McIvor, J., O'Brien, M., Wright, C.F.; Industrial Relations Journal, Feb. 2019; https://onlinelibrary.wiley.com/doi/full/10.1111/irj.12238, Retrieved April 12, 2020.

[2] Intertek, Restricted Substances List (RSL) Services, 2019, https://www.intertek.com/green/restricted-substances/list/, Retrieved April 12, 2020.

[3] European Commission, Product Environmental Footprints (PEF) Guide, https://ec.europa.eu/environment/eussd/pdf/footprint/PEF%20methodology%20final%20draft.pdf, Retrieved April 12, 2020.

[4] International Facility Management Organization, What is Facility Management?, https://www.ifma.org/about/what-is-facility-management, Retrieved April 12, 2020.

[5] Weele, A.J.V., Procurement and Supply Chain Management: Analysis, Strategy, Planning and Practice (5th ed.). Andover: Cengage Learning. ISBN 978-1-4080-1896-5., 2010.

[6] Prasad, S., Procurement organizations have the power to reduce greenhouse gas emissions, 2017, https://www.beroeinc.com/blog/procurement-organizations-power-greenhouse-gas-emissions/, Retrieved April 12, 2020.

[7] Halbheer, D., Paris, H.E.C., The Role of Marketing in Climate Change: Carbon Footprinting and Pricing, https://www.hec.edu/en/knowledge/articles/role-marketing-climate-change-carbon-footprinting-and-pricing, Retrieved April 12, 2020.

[8] Rouse Magazine, Holly White, General Counsel, can you afford to ignore climate change?, https://www.rouse.com/magazine/news/general-counsel-can-you-afford-to-ignore-climate-change/, 2018, Retrieved April 10, 2020.

[9] Kuo, T.. Simulation of purchase or rental decision-making based on product service system. International Journal of Advanced Manufacturing Technology. 52. 1239–1249. 10.1007/s00170-010-2768-2, 2011.

[10] Product Sustainability Forum, Steps to be taken for Eco-Design of a material product http://www.wrap.org.uk/sites/files/wrap/Embedding%20sustainability%20in%20design%20%20-%20final%20v1.pdf, Retrieved April 10, 2020.

[11] Rouse Magazine, Holly White, General Counsel, can you afford to ignore climate change?, https://www.rouse.com/magazine/news/general-counsel-can-you-afford-to-ignore-climate-change/, 2018, Retrieved April 10, 2020.

[12] Reducing carbon emissions through employee participation: Evidence from Australia; Markey, R., McIvor, J., O'Brien, M., Wright, C.F.; Industrial Relations Journal, Feb. 2019; https://onlinelibrary.wiley.com/doi/full/10.1111/irj.12238, Retrieved April 10, 2020.

Chapter 10 Audits

10.1 General

In addition to its own agreed processes, a company must also ensure the processes of its partners in order to comply with the quality agreements with its own customers. The Green Procurement Standards mentioned above describe the environmental requirements that have to be fulfilled by clients and customers. Audits are carried out to check compliance with the agreements – also to obtain an (independent) view of the environmental aspects of the activities. As the "Environmental Auditing Technical Report of UNEP/IEO" (14) states, an environmental audit (EA) is "a management tool that involves a systematic, documented, periodic and objective assessment of the performance of the environmental organisation, management and equipment, with the objective of contributing to the protection of the environment through:
- Facilitating the management and control of environmental practices,
- assessing compliance with company policy, which includes meeting legal requirements".

The EU offers a system for environmental management and auditing (EMAS). It is an environmental management system and formalised as a regulation [1]. This system is currently a still voluntary tool to improve environmental performance and support efforts to ensure compliance with environmental legislation. According to this regulation (1), the priorities of a company's environmental performance include criteria as the ones in Table 10.1

Table 10.1: Priorities of a company's environmental performance. It should be noted that resource efficiency is therefore treated under different aspects (according to [1]).

– The efficient use of natural resources (e.g. energy, water, materials)
– Waste prevention
– Recycling and reuse
– Green procurement
– Sustainable mobility

The FAO, with reference to the International Chamber of Commerce (ICC), provides a useful description of how such an audit can be carried out [2]. Here the logic described in [2] is followed. The audit can focus on organisation, emissions, compliance with standards and regulations, maintenance, safety, material balance, training, external contractors, etc. It is important to note that the audit system represents a cycle. Its periodicity must be pre-determined and experience shows that the periodicity of the site audit phase should not exceed three years.

https://doi.org/10.1515/9783110767308-010

The International Chamber of Commerce presents the different steps of an EA as follows (Table 10.2):

Table 10.2: Steps for an environmental audit in an organisation [2]. It should be noted that the audit can focus on selected topics.

1. Pre-audit activities: These include selecting and scheduling the facility to be audited, selecting the audit team, contacting the facility and planning the audit.
2. On-site activities: which are divided into 5 steps: Understanding internal controls, Assessing the internal controls Collecting audit evidence Evaluation of audit results Reporting the results to the facility
3. Post-audit activities: These include preparation of a draft report, preparation of a final report, preparation and implementation of an action plan, and monitoring of the action plan.

The ICC presents the content to be delivered by an EA [2]. At a high level, the EA results in a report that could be structured as follows (Table 10.3).

Table 10.3: Structure of the report after the environmental assessment.

1. Executive summary: Should enable anyone (professional or not) to understand the various impacts of the project and be informed about the alternatives chosen and the mitigation measures to be taken.
2. Project description and legal and administrative framework: A brief description of the project with all off-site extensions and their interaction with natural and social components, including all regulations implemented as part of the environmental audit, should be provided here.
3. Scoping and screening: Allows those responsible for environmental protection (ministries, borrowers, donors, NGOs, associations, residents, etc.) to outline to the investor the limitations of the environmental audit in terms of time, space and the type of impacts to be considered, and to show the alternatives.
4. Description of the existing environment: Provides accurate data relevant to the site and describes: intended use, quality, physical, biological, social and economic conditions. This description must include other existing or planned developments. The judicious use of maps, diagrams, drawings, etc. is important for a better understanding of the situation. Important data gaps and uncertainties must be identified.

Table 10.3 (continued)

5. Analysis of alternatives and basis for selection of the proposed alternative: The project description is complemented by a detailed description of the different choices regarding processes, location and all alternatives that the investor has considered for better protection of the environment and the affected population. A comparison of these different alternatives in terms of their potential impacts and a cost-benefit analysis is required. The basis on which each alternative was chosen must be stated.
6. Environmental aspects of the project: Once the project has been defined and all alternatives have been thoroughly investigated, this section presents the environmental aspects of the final project. Each area of positive or negative impact must be defined in terms of its size, reversibility, period of occurrence and type (primary, secondary, etc.). At this stage, it is important to outline in detail the different phases of the project and address all environmental impacts associated with each phase. Any drastic negative impacts that cannot be eliminated must be identified and minimally mitigation measures must be proposed.
7. Mitigation measures: For all remaining negative impacts, mitigation measures must be proposed and scheduled in a timely manner. These measures must be both technically and economically realistic. The efficiency of each measure to reduce significant negative impacts to an acceptable level must be assessed. An estimate of the investment required is necessary at this stage to verify the feasibility of the proposed measures.
8. Environmental management and training and environmental monitoring plan: In order to try to avoid environmental accidents, it is necessary to prepare a document defining the role of each person or group in the environmental management team of the future enterprise, as well as the monitoring and training procedures that will be carried out to improve the skills of employees and workers. These documents will of course need to be updated after the plant is built.
9. Annexes: All documents needed to understand the chosen methodology, the references, the meetings with government bodies, scientists, managers, concerned groups, the names and qualifications of the authors of the study must appear under this heading.

According to FAO [2], an environmental audit with follow-up report provides benefits such as given in Table 10.4:

Table 10.4: Benefits of environmental auditing according to [2].

- Facilitating comparison and exchange of information between farms or facilities
- Raising staff awareness of environmental policies and responsibilities
- Identifying cost savings, including those resulting from waste minimisation
- Evaluate training programmes and provide data to support staff training
- Providing an information base for use in emergency response
- Ensuring an adequate, up-to-date environmental database for internal management awareness and decision making in relation to plant modifications, new plans, etc.
- Enabling management to recognise good environmental performance

Table 10.4 (continued)

– Supporting relations with authorities by convincing them that full and effective audits are being carried out by informing them of the nature of the approach
– Facilitating the obtaining of insurance cover for liability for environmental damage

10.2 Questionnaires

Audits inevitably go hand in hand with asking relevant questions. Especially when dealing with partner organisations, it is important to note that the audits have been agreed at the time of a contractual arrangement that is subject to audits and, in the best case, have been gone through in advance with the partners and the own organisation. Several organisations and companies have developed examples of questionnaires assessing the relevant parameters in an organisation. As structured above, the questions must aim to understand the product and its environmental footprint and provide the relevant information about the organisation involved and its objectives, processes and actions.

10.2.1 General

Ultimately, the scopes of the GHG Protocol (the criteria to be determined can be found in the lists in Chapter 5) are best suited for organising the process of determining emissions and their causes. It proves extremely helpful to name one responsible individual per sub-scope. Administrative framework conditions must be recorded for all parameters:
- Description of the calculation method
- Annual statement or invoice
- Installation number, if available
- Calculations or assumptions
- Where the data is stored
- Responsibility for the (equipment and) data

The relevant parameters are based on the questions on scopes as used in Chapter 5.

Often the determination of product-related emissions requires an industry-specific look into what is produced and the associated production methods. Above, in the example for analogue newspapers, it was explained that the paper and its transport, the ink used, the waste (packaging and waste paper) and the printing plates account for well over 80% of all emissions in the production of a printed product (see examples above). Accordingly, much emphasis is placed on the accurate determination of this parameter when determining emissions in the printing industry.

In addition, there are the peculiarities of transporting the materials to and from the production site which can be assessed using questions like the one displayed in Table 10.5:

Table 10.5: Questionnaire for the emission footprint according to [3].

Transportation of products to and from sub supplier (purchased transportation)*	
*Transportation of products in own vehicles must be included as fuel consumption under "Purchased fuel for burning in the company's own or leased vehicles" (page 1).	
Yearly amount of transported products in truck with load weight capacity < 3,5–16 tons	kg
Average distance	km
Yearly amount of transported products in truck with load weight capacity of 16–32 tons	kg
Average distance	km
Yearly amount of transported products in truck with load weight > 32 tons	kg
Average distance	km
Description of calculation method: – Yearly statement or invoice – Where are the data filed? – Calculations or assumptions – Responsible for data	
Transportation of products to customers (purchased transportation)*	
*Transportation of products in own vehicles must be included as fuel consumption under "Purchased fuel for burning in the company's own or leased vehicles" (page 1).	
Yearly amount of transported products in truck with load weight capacity < 3,5–16	kg
Average distance	km
Yearly amount of transported products in truck with load weight capacity of 16–32 tons	kg
Average distance	km
Yearly amount of transported products in truck with load weight > 32 tons	kg
Average distance	km
Description of calculation method: – Yearly statement or invoice – Where are the data filed? – Calculations or assumptions – Responsible for data	

As described, this type of questionnaire is a more specific industry- and product-related formulation of the GHG scopes questionnaire: The cut must be such that the questions build on prior knowledge of industry and product characteristics. Often, a good internet search will provide templates and initial ideas that fit the particular requirements.

The conversion of the data contained in this table to CO_2 footprints then takes place using the conversion factors deposited in the literature – i.e. for example x kg of paint corresponds to y kg of CO_2 depending on the type of paint or x kg of plastic corresponds to y kg of CO_2 depending on the type of polymer and degree of recycling.

10.3 Reporting

For reporting purposes, the sequence of project steps can be followed, as given, for example, by Motzer and his guide to corporate carbon footprint reporting (4):

1. Report system boundaries and develop greenhouse gas inventory.

Report the system boundaries of the calculated GHG inventory.
- Relevant production sites (Company)
- The more holistic, the higher the potential for GHG emissions and thus for cost reduction
- Activities that result in GHG emissions – especially those relevant to the CO_2
- Inventory depending on the sector, product range, degree of innovation, etc.
- Extent to which these activities can be controlled and influenced
- Scope and weighting
- Underlying sources and types of activity data

2. Determination of consumption values

Robust information or estimation of consumption data including the company's environmental footprint:
- Quantified energy and material consumption plus other activities of the operation identified as significant
- Depending on the boundary conditions and local situation: activities outside the organisation (e.g. third party transport, waste treatment, other relevant activities in the supply chain)
- Assessment of the impact of missing, incomplete or insufficient data
- Plausibility check of the collected data, if necessary with the help of industry benchmarks

3. Calculation of the Environmental Footprint
- If used: name and source of the emissions calculator that complies with the standard
- Breakdown of results in terms of scope, operation or process (depending on the granularity of the system boundaries set)

4. Development and communication of a footprint strategy, reduction targets and indicators
- Material and energy strategy
- Reflection reflected in corporate strategy, including risks and opportunities
- GHG reduction targets and means to track them
- Identification of hot spots (and low-hanging fruit)
- Realistic but challenging reduction targets and means of tracking them
- Reconciliation of energy and material impacts in relation to processes, products and services to track performance improvements and cost effects

References

[1] European Commission, REGULATION (EC) No 1221/2009 OF THE EUROPEAN PARLIAMENT AND OF THE COUNCIL, on the voluntary participation by organisations in a community eco-management and audit scheme (EMAS), repealing Regulation (EC) No 761/2001 and Commission Decisions 2001/681/EC and 2006/193/EC, 2009 (=EU Regulation (1221/2009/EC)), https://eur-lex.europa.eu/legal-content/en/ALL/?uri=CELEX%3A32009R1221, Retrieved April 10, 2020

[2] FAO, Environmental Impact Assessment (EIA) and Environmental Auditing (EA), http://www.fao.org/3/v9933e/v9933e02.htm, Retrieved April 10, 2020

[3] ClimateCalc, Tools for collection of data, https://www.climatecalc.eu/950.aspx, Retrieved April 10, 2020

[4] Thomas Motzer, 5 Steps to Successful Corporate Carbon Footprint and CDP Reporting, Thinkstep, 2019 https://www.thinkstep.com/blog/5-steps-successful-corporate-carbon-footprint-and-cdp-reporting), Retrieved April 10, 2020

Chapter 11 Information and transparency

Despite the trend towards low-emission and material-saving products, the status of green developments by companies is often unclear. Green development assessment indicators are needed for public and private stakeholders, especially as companies have often issued detailed green development guidelines and made them available to the public. Reporting is an important tool to foster communication between governments, the public, companies and stakeholders, and to assess and provide guidance on progress towards green development.

As discussed in the introduction, the move towards a more sustainable economy is not really new. It has also been shown that, for example in the EU, the legal framework is extensive and covers many of the necessary areas. Nevertheless, the transformation process towards a low-emission society is not yet complete and communication is part of this process. Several target groups for communication need to be addressed with consistent messages (Table 11.1):

Table 11.1: Target groups for communication on environmental activities.

– Employees
– The general public
– Customers
– Shareholders
– Partners
– Government agencies
etc.

These target groups are well known in terms of their mind set and the type of ties held. All of them need specific channels and, as mentioned above, reliable, reproducible and clear messages using appropriate channels. In particular, the channel chosen can be crucial.

While for a client or a partner a phone call may be considered inappropriate if the message is significant, channels such as social media and press releases are a good choice for communicating to a wider audience. These means of communication are well established and in some cases even subject to legal regulations. Nevertheless, cultivating a company's image and creating a brand and loyalty to it goes hand in hand with addressing business challenges, such as an environmental strategy.

https://doi.org/10.1515/9783110767308-011

11.1 Communication with employees

Commitment to sustainability and taking action are crucial elements of corporate activities. However, without convincing and involving employees, such initiatives have little chance of success. Effective communication with employees creates transparency, or at least a sense of it. Communication creates a sense of trust – employees want to be part of the whole and assume that they are, provided they are communicated with clearly. Poor communication can cause a rift between companies and their leadership and thus a disruption in the flow of action. There is no single roadmap for engaging employees – different solutions work in different environments, companies have different cultures, and the maturity of the company's sustainability strategy is also a critical factor – but much is already being done by companies to get employees on board. Such solutions create initiatives, results and thus messages to be communicated. As seen in Table 11.2 below, these can be broadly divided into four categories:

Table 11.2: Means to get employees on board to identify and implement solutions for a reduced environmental footprint (see also [1]).

Segmentation and survey Segmentation into employee characteristics and affiliation to parts of the company or hierarchical levels helps to structure interest in sustainability and capture it according to a structure. It helps employers to formulate the right sustainability message for the right group. It also forces companies to consider employees with very different characteristics of attitude (positive or negative to the measures being considered).
Promotion and team building In seeking ecologically motivated behaviour change, surveys, focus groups and sampling are used to identify the right actions and communication channels, and in particular become part of jobs and job descriptions. Ideally, this is only one aspect of a comprehensive approach that aims to close any gaps that may exist between existing attitudes and the desired target behaviour.
Dedicated teams Dedicated teams and employees are an integral part of corporate sustainability strategies. Qualified employees with a stake in success can take the lead and monitor measures and target achievement (such as recycling or energy consumption) according to specifications. It is supportive if these teams and employees are given a defined proportion of their time for sustainability practices.
Volunteering Volunteering is another approach that gives employees the opportunity to make a small, individual decision that has a positive impact on their community and/or the company. The simpler or more insightful the practical action, the more likely it is that employees will participate in implementing sustainability strategies.

Susan McPherson [2] proposed the rules for best practice in sustainability communication with employees (Table 11.2):

Table 11.3: Best practice rules for sustainability communication according to McPherson [2].

1) Reach employees where they are:
   In break rooms, conference rooms, washrooms, car parks and online. Reach employees outside of actual working hours.
2) Integrate sustainability into working life from the start:
   From the first day a new employee starts, they should be given a full overview of the company's sustainability plans and efforts.
3) Show progress and best practices:
   Regularly inform employees about your company's results. By keeping employees informed, they can be involved in sustainability discussions and planning. As sustainable behaviour is part of an ongoing process and cannot be completed, it is important to acknowledge improvements to encourage motivation.
4) Use technology effectively and appropriately:
   Choose your tool depending on the desired outcome and the target group.
5) Coordinate with HR/internal communications/PR to ensure all messages (internal and external) are consistent.
6) Offer meaningful incentives to encourage interest, participation and feedback:
   Companies can provide dedicated workstations as well as a monthly "prize" to award employees for creative ways to recycle and reduce emissions.
7) Recognise key ambassadors and thought leaders:
   Find the employees who are actively interested in sustainability and give them the opportunity to take a leadership role.

11.2 Communication with the public

Communication with the public aims at supporting the company's environmental strategies and measures to implement them, raising public awareness about ongoing, planned and completed activities and achievements, the company's values and priorities and efforts to overcome challenges. Communication about the context – such as local air quality, industrial emissions, water and oceans, the global context as well as the effects of the company's own products and generated waste on the environment, but also nature conservation, policy initiatives and infrastructure development falls into this category.

Communication activities to the general public are intended to both shape understanding of environmental needs and improve knowledge of implemented values and policies that lead to responsible growth, human health and well-being. This, in turn, is intended to build support for and improve the implementation of the company's policies, as well as demonstrate the company's added value to the community. The main objectives are to:

- Inform (and where appropriate engage) a wide range of audiences on environmental priorities;
- Create awareness that the environmental policy (EP) is associated with a wide range of benefits for the company and their customers, and that the approach chosen and followed can demonstrate positive results;
- Create and maintain a positive public perception.
- Present the progressive and standard-setting role of the company.

Communication activities aim to show how the actions contribute to environmental conservation, health and sustainable well-being, without neglecting the economic objectives and needs of a company. An appropriate reporting should include at least- the sustainable business model as an integral part of its strategic planning and existing code of conduct. A minimal content is suggested in table 11.4 below:

Table 11.4: Minimum content of reports showing how environmental policy contributes to health and well-being without neglecting a company's economic objectives and needs.

– what a commercially successful organisation uses, does, produces and delivers in terms of desired environmental outcomes
– environmental planning considerations, support to top management in institutionalising environmental concerns into environmental corporate structures and organisational features, environmental governance activities, environmental auditing, external validation or certification of environmental programmes and forms of corporate environmental reporting.
– reporting on the organisational and product aspect indicators, highlighting sustainability, economic and social indicators (SDGs).
– key internal factors influencing environmental matters, policies and standards followed (e.g. quality (ISO 9001), environmental (ISO 14001) and occupational health and safety (OHSAS 18001)), legislation adopted, environmental controls, external validations or certifications of environmental programmes and forms of corporate environmental disclosure.
– Environmental benchmarks achieved

Reports can be assessed to evaluate their content. The assessment can be based on indices and environmental indicators proposed, for example, by the Global Reporting Initiative (GRI) to review the information published in corporate social responsibility reports.

11.3 Communication with customers

Besides what a company produces, customers are interested in what a company stands for. Therefore, being transparent with customers helps to build trust. A reliable "face" to the customer and partner is important as this improves customer loyalty. On the other hand, customers may feel they are not receiving the best possible service if staff interacting with customers have poor communication skills or are not familiar with the activities or policies – because they are not clearly communicated or they mishandle them and give false information to the customer. As indicated above, positive, effective external communication starts internally. If the company clearly communicates its values and goals, employees are better able to reflect these qualities when representing the company. Communication with customers is through publications, sales experts and their management during visits and exhibitions.

Communication through sales

It is not uncommon, for example, for even larger companies to require their suppliers to agree to their code of conduct, but beyond that, environmental issues are not talked about in product communications or by sales people. Nevertheless, companies can advertise more effectively if they present themselves in the context of sustainability and their own activities in the field. Sustainability work can be used as a competitive advantage, but only if the advantages of being a sustainable supplier are made clear to customers (stability and a sense of responsibility). If customers do not actively ask about environmental measures, this does not necessarily mean that customers would not recognise the value if it were addressed. Because the issue is problematic and new to some, there is a certain reluctance on the part of sales to raise the issue with customers. If sustainability work is to be used as a competitive advantage, sales must be able to initiate a discussion about the benefits with customers – or at least address the issue with confidence. It is often sufficient for sales to use a product- and company-related talk track.

The product-related talk thread should refer to environmental design principles that represent the state of the art at best, and can cite literature to emphasise relevance and gain credibility. In general, what can be addressed for a product is what was focused on in the product design phase – these are in particular the sustainability strategies pursued – see Table 11.5

Table 11.5: Possible content of the product-related address for distribution.

1	Supply chain
2	Reduction of energy consumption during production and use
3	Parts and materials strategy, which includes Composites and variations in alloys to increase the value and recycling rate of the product's parts. Increase the recycling rate of the product's parts that enables reuse and refurbishment, and the use of special design measures and services to extend the service life (of the system itself and the parts) and to allow a second life (or more) after remanufacturing
4	Ability of the system to use recycled and recyclable material, if possible industrially compostable materials.

One can often find references to carbon and material footprints of competing or similar products in the literature on comparable products. Even if these do not fit exactly and if the assessment is provided together with the quantitative results, they often give an indication of the main polluters and thus show ways in which manufacturers and customers can influence which parameters in order to achieve a reduction of the environmental footprint. Meanwhile professional customers are often aware of this information – so proactive research and communication helps – as does suggesting simple measures on the customer side such as using renewable energy sources or even allowing remote preventive maintenance to extend the life of an installation.

In addition to communicating about the company's own environmental initiatives, it is important to pay attention to the activities of partners who supply parts and materials in external communications, for example, about the established Green Procurement Standards that outline the environmental requirements for suppliers – but also to what extent, if a supplier is found to have a negative impact on the environment, corrective action is demanded and the status of improvements made is checked.

The communication guide should refer to the production characteristics of the product and relevant site, the main teams working together at the site should be mentioned to show how sustainable cooperation works. The site-specific guide reflects whether or how the organisation has changed to use alternative energy sources – the energy supply – for example. In this type of communication, detailed indicators such as in the example of energy supply for the percentage of electricity generated from renewable sources, heating covered by geothermal energy and details on cooling, for example, to what extent cooling is done with groundwater via a central cooling system, are supportive.

Pre-structured talk tracks also prove helpful for inexperienced sales staff. For more detailed information, an expert can be consulted. However, by far, not all salespeople today experience the pressure from customers regarding sustainability.

In this context, it is important to know the customers' sustainability work and goals. If the customer does not raise environmental issues in conversations, it is important that salespeople are able to introduce the topic and communicate the added value of environmental measures to the customer's supply chain. Also in terms of environmental footprint, a supplier needs to guarantee a certain level of quality in terms of sustainability and environmental measures and take on some of the client's responsibility in terms of sustainability towards its customers and stakeholders. Highlighting environmental benefits is worthless to the customer if the product as such does not meet the required functional and technical performance. It is useful to communicate environmental performance in a way that makes it clear that it is relevant to the customer and linked to the functional or technical performance prioritised in the application.

To discuss an example:
When sales specialists communicate with client-side procurement or maintenance departments, they may be less interested in environmental savings if, for example, the cost of unexpected business interruptions is higher than the cost of the energy that an energy-efficient product can potentially save.

Often environmental aspects are also not yet on the agenda and targets of the client side and its management, or attention does not go beyond existing legislation. In these cases, it is unlikely that environmental factors will be taken into account in procurement decisions. Nevertheless, in the medium term, salespeople need to consider environmental performance in their communication with customers, even if at first glance it does not add value to the product offering for customers. This is positive, as a supplier can achieve competitive advantages in the long term through proactive and adequately communicated sustainability work.

Obviously, sales need environmental education to communicate confidently, as they need to feel comfortable approaching customers about sustainability. There should be some pressure to participate in such training. If not the whole, then a large part of an organisation needs to be involved and convinced, as sustainability efforts often require the cooperation of the whole organisation to achieve sustainability goals. Therefore, internal marketing for sustainability issues is important.

Communication at trade fairs

National and international trade fairs are important to present companies and products. Environmental sustainability management and the message that products and services have been designed with the environment in mind must be part of the message conveyed at these events. Beyond talk tracks, special areas and easy-to-understand exhibits help to show what has been done in design to offer environmentally friendly solutions. In times of change, a certain educational aspect of such displays and engaging talks with experts should not be underestimated as an orienting measure. As indicated, a small environmental footprint is irrelevant if the required benefits are not delivered, but if they are, it can make the difference to a deal, especially when legislation and auditing demand more sustainable solutions.

It cannot be stressed enough that transparent communication, even if negative at first sight, is helpful. Providing background information and thus providing customers with messages that support the company's mission has an impact on the company's reputation.

Communication via publications

Publications, especially those by internal experts in general and specialised publications, influence the brand image in order to be perceived as a sustainable supplier. It is helpful to combine content-based insights with a well-communicated product-specific environmental performance in order to improve the respective brand image. A suitable reputation as a sustainable supplier is a necessity to successfully communicate the benefits of product-specific environmental performance. The messages can be enhanced by including certain information in the communication as displayed in Table 11.6 below:

Table 11.6: Messages to increase reputation as a sustainable supplier.

(1)	Commitment to continuous improvement of environmental performance with a focus on implemented processes and their auditing.
(2)	Reference to labels and third-party certifications for defined and named standards.
(3)	Reference to existing partnerships and their supply chain involvement, as suppliers are in turn held accountable for their supply chain and sustainability requirements are transferred backwards in the supply chain.
(4)	towards particularly environmentally conscious customers: Offer communication of life cycle assessments in a complex and comprehensive form.
(5)	towards environmentally unaware customers:
Communicate easy-to-understand information, but still mention the methods used to obtain the basic environmental footprint information, as these customers also want figures with units.	

11.4 Communication with share holders

Communication with partners and shareholders is about transparent information about the status, the successes in the reporting period and the existing risks. In some cases, the communication of such information is required by law or demanded by institutional investors. A focus on companies with strong environmental, social and governance (ESG) performance in capital investment decisions is not morally tainted, but rather due to the positive performance and more secure corporate development of companies that focus on these aspects. For example, [3] concludes that companies with a better ESG record than their peers had higher three-year returns, were more likely to become high-quality stocks, were less likely to have large price declines, and were less likely to go bankrupt. Firms with good ratings on sustainability issues [4] perform significantly better than firms with poor ratings on this issue area. It is therefore natural that manufacturers would like to be seen in this context.

According to the Boston Consulting Group, the discrepancy between what investor relations officers and investors consider important is remarkable [11]:

While 75% of investors state that sustainability performance is essential for investment decisions, only half of investor relations officers consider sustainability important for the company. According to the same study, this significant gap is not due to investor relations not knowing the big picture, but to the fact that they do not receive guidance on sustainability communication with shareholders. Furthermore, operational managers and sustainability experts typically have very limited understanding of the investor community and little interaction with shareholders. As official spokespersons for the company, it is the company management that has access to investors, not the sustainability managers. For this reason, investors do not ask sustainability professionals for information on related initiatives – rather, they turn to investor relations' who, if not advised and trained, are not able to lead the necessary discussion. Almost 80% of IR managers say they do not include sustainability aspects in their investor discussions.

Moreover, when investors approach managers, they often get lost in the different terminologies used for sustainability activities to such an extent that even experienced sustainability managers are lost in the resulting conversations. To prevent this uncertainty, investors can use more colloquial language or express themselves in more general terms such as: "Tell me, how is the plant doing? How is the energy consumption? How are the fuel prices affecting it?", which is also not conducive to getting concrete answers. It is helpful to test the relevant talks in advance in order to have suitable talk tracks and documents at hand.

Close cooperation between the IR department and the sustainability function is very beneficial in this context:

Sustainability management provides information and helps respond to specific investor queries as they arise. It contributes when analysts or investors ask for sustainability-related information, jointly develops an equity talk track for the company's sustainability initiatives and shows how sustainability investments contribute to investors' concerns about efficiency, risk and growth.

According to Eccles, Ioannou and Serafeim [3], sustainable investing from the investors' perspective comprises a menu of strategies that can also be used in combination and should be addressed accordingly in communication (see Table 11.7)

Table 11.7: Investor menu of sustainable investing strategies.

Approach	Strategy	Main characteristics
Negative selection	Negative/ exclusionary screening	eliminating companies in industries or countries deemed objectionable
	Norms-based screening	eliminating companies that violate some set of norms, such as the Ten Principles of the UN Global Compact
Positive selection	ESG integration	including environmental, social, and governance (ESG) factors in fundamental analysis
In sequence of depth of focus	Impact investing	looking for companies that make a positive impact on an ESG issue while still earning a market return

Table 11.7 (continued)

Approach	Strategy	Main characteristics
By focal areas in portfolio	Positive/best-in-class screening	selecting companies with especially strong ESG performance e.g., building a portfolio of companies that have a carbon footprint 50% + smaller than benchmarks and have 50% less exposure to "stranded assets"
	Active ownership	engaging deeply with portfolio companies
	Sustainability-themed investing	such as in a fund focused on access to clean water or renewable energy

Companies that focus on material sustainability issues tend to perform better than those that do not, according to the Sustainability Accounting Standards Board (SASB). SASB has identified material ESG issues for 77 industries in its classification system [6]. For example, for companies in food distribution and retail, material issues include greenhouse gas emissions, energy management, access and affordability, fair labour practices, and fair marketing and advertising. For internet and media services, the corresponding list includes energy management, water and wastewater management, data security and customer protection, diversity and inclusion, and competitive behaviour.

11.5 Communication with partners

Suppliers have a significant impact on the success of a company through their influence on costs, supply chain and environmental performance. Therefore, agreements with partners on ESG goals also have a strong impact on success. In addition, audits are important because while working with reliable, high-quality suppliers can help to grow, unreliable suppliers can create bottlenecks in a company's workflow and introduce social and environmental risks and cause supply stoppages. To manufacture and distribute their products, companies rely on an often global network of suppliers. From the perspective of the customer and the legislator, however, the responsibility for the product lies with the party putting it into circulation. He is liable from the customer's point of view. In order to meet the growing demands of customers, companies must therefore work closely with their suppliers, also in terms of liability. To achieve best sustainability practices, communication with suppliers must be cooperative and transparent, with stated common goals and clear definitions for measuring efforts.

Appropriate communication must be embedded in supplier relationship management (SRM). SRM was first introduced in 1983 by McKinsey consultant Peter Kraljic [7]. It is essential for incorporating ESG due diligence strategies throughout the supply chain. The SRM concept is based on the idea that suppliers are business partners, and as such this partnership must be strategic and growth-oriented, rather than reactive. The goal of identifying, selecting and maintaining a relationship with a supplier is to ensure that suppliers who are cost effective, easy to work with and have moderate or low social and environmental risks are commercially viable to maintain and maximize the value of the relationship – it is not just profit maximisation that counts. Obviously, the partner's supply chain, including ESG-related issues, also counts when choosing a partnership. Therefore, it is necessary to communicate and explain one's corporate strategy in terms of ESG policies and objectives upfront and when changes occur. Furthermore, it must be clear how this strategy is anchored in the signed contractual agreements including corresponding audits and "end of business" clauses.

This approach ensures the company's competitive advantage in the long term and helps companies to become more effective in their due diligence. How a partner handles the communication and resolution of potential ESG crisis is an integral part of quality management.

Producers, manufacturers, distributors, logistics providers and other parties have a vested interest in getting their products to market quickly, safely and under the best possible conditions. There is considerable pressure on managers to ensure responsible supply chain behaviour [7] – in particular to streamline relationships with their partners and reduce the risks of their chain. The objectives, in addition to maximising the overlap of interest between business value and social and environmental value, are to focus on initiatives that improve business value and the environment, business value and the local economy/society. If all tree parameters – profitability, environmental issues and the needs of local economies or societies are met [7] speaks about the "triple advantage" in which socio-environmental and business values can be met by protecting the environment plus a more stable and vital local economy and society.

The World Economic Forum Report [7] shows that significant potential benefits can be achieved by implementing these "Triple Advantage" practices (modelled first on the consumer goods supply chain) (see Table 11.8)

Table 11.8: Benefits from responsible supply chain behaviour according to [7].

– Profitability:	increased sales
Supply chain:	cost reduction
	Brand equity enhancement
Business risk:	significant reduction of turn over
– Local development and social: standards (wages, working conditions)	local economic growth and improved labour
– Environment:	Reduction of carbon dioxide emissions through reductions in all scopes

The report [7] identified a comprehensive set of 31 good practices, supported by industry examples, to guide companies in implementing measures to improve the triple bottom line.

11.6 Corporate sustainability report

A company's sustainability report is the main publication in the context of sustainability. These reports aim to inform all communication audiences about environmental strategies, goals and activities. In 2018, around 80% of the S&P 500 produced a sustainability report [8]. Indirectly, the source [8], but also the Sustainability Assurance Standards Board [5], the IIRC international Integrated Reporting Council [9] and the GRI Sustainability Reporting Standards and Guidelines (www.globalreporting.org/information/sustainability-reporting/Pages/gri-standards.aspx) give a clear idea of what should be part of a corporate sustainability report:

Table 11.9: Minimum content of a sustainability report.

– Quantified and time-bound environmental targets
– Environmental performance indicators (quantified measures that are comparable from year to year)
– Environmental performance data (as verified externally) such as greenhouse gas emissions
– Targets and metrics (including injury and accident rates)
– Externally verified data
– Information on standards followed, legislation considered, and external assurances and audits on these.
– Sustainability reporting should follow a published and proven framework (GRI, SASB or, less easily communicated: an industry-specific model)
– Voluntary sustainability information

The sustainability report starts with the foundation laid by the other means of communication. It should be accompanied by a letter from the CEO showing the direction of the leadership and clarifying how the sustainability efforts fit into a bigger picture, for example, by linking it to related EU initiatives.

11.7 Environmental Policy

Tearfund, a US organisation involved in humanitarian and development work, has developed a model EP for an imaginary development organisation [10], which can serve as an example to display how it was developed and what it contains:

In general, an EP is a commitment made by an organisation and a statement of what the company considers to be the most important activities to implement that commitment. With this paper, a company acknowledges its impact on the environment and combines this with the desire to minimise this impact. The wording of the EP enables the company to communicate environmental objectives to employees and the outside public (partners, customers, stakeholders and other companies). Through the EP, the organisation's management communicates in a binding way its attitude and the resulting actions and appreciation of the environment combined with the will to take care of the environment in order to preserve it for future generations. The document should be worded by experts in environmental issues but also by people who know the company – only such people can ensure that the policy is appropriate for the organisation and its products and network. It may be beneficial to bring in an external consultant to work with staff on developing a EP. All staff need to take ownership of the agreed policy as they are the people who will put it into practice. GRI and sustainability report [10] offers practical tips for creating a EP:

Ensure that the policy is appropriate for the organisation.

- The objectives of the EP relate to the activities of the organisation.
- The objectives should be realistic and achievable.

Encourage staff to read the policy, take ownership of it and so on.

The policy shall be

- short (e.g., one or two pages)
- generally understandable (in terms of language and translated if necessary)
- easily accessible and visible and the policy is implemented
- Policy is signed and promoted by management and is therefore taken seriously.
- Managers behave in an exemplary manner and as pioneers.
- Awareness raising measures are implemented.

The EP should include a clear commitment to continuous improvement and specific actions referred to in the awareness raising, participation and training of staff on environmental issues.

The EP should also include informed references to quick fixes (such as double-sided printing) and longer-term solutions (such as reducing the number of flights or planting trees), as well as a commitment to help employees understand and act on environmental sustainability issues. Simple examples are displayed in Table 11.10:

Table 11.10: Content of the environmental policy according to GRI and sustainability report [10].

Travel The company encourages its employees to use environmentally friendly forms of mobility, that is, walking, cycling, using public transport and minimising the use of cars and air travel.
Office practice The company saves energy wherever possible in its buildings, with particular attention to heating, lighting, ventilation and office equipment. Water is used efficiently and carefully. The use of renewable energy is encouraged.
Waste management The company strives to reduce, reuse and recycle waste as much as possible. This includes recycling waste paper, cans, plastics and CDs. The organisation is also committed to reducing the generation of waste. To achieve this, the purchase and use of items made of plastic and polystyrene that are not biodegradable is discouraged.
Office supplies management The company sources recycled office supplies wherever possible and uses printers and inks that take into account environmental impacts. Paper consumption is monitored and ongoing measures are taken to minimise the amount consumed. Recycling of paper products is actively promoted. Whenever possible, electronic communication is preferred to paper.
Procurement and trading Environmental factors should always be considered in the procurement activities of The Company. Although cost considerations are important, preference should be given to local suppliers who take good environmental practices and sustainability seriously.
Policy management and monitoring The organisation will appoint a staff member as a part-time environmental officer who will be responsible for developing and reporting on an annual action plan.
Environmental audit The organisation will conduct a full environmental audit once a year and make the results openly available to others.

References

[1] Rodolphe d'Arjuzon, Engaging Employees on Sustainability, https://www.theguardian.com/sustainable-business/employees-engaged-sustainability-how, Retrieved April 10, 2020

[2] Susan McPherson, Fenton, quoted in Ellen Weinreb, Communication Strategies to Engage Employees in Sustainability, https://www.greenbiz.com/blog/2011/10/12/10-communication-strategies-engage-employees-sustainability October, 2011, Retrieved April 10, 2020

[3] Robert G. Eccles, Ioannis Ioannou, George Serafeim, The Impact of Corporate Sustainability on Organizational Processes and Performance, HomeManagement Science Vol. 60, No. 11 https://pubsonline.informs.org/doi/10.1287/mnsc., 2014

[4] Mozaffar Khan, George Serafeim, and Aaron Yoon Corporate Sustainability: First Evidence on M ateriality, The Accounting Review, Vol. 91, No. 6, pp. 1697–1724, 2015 (2017) https://papers.ssrn.com/sol3/papers.cfm?abstract_id=2575912, Retrieved April 10, 2020

[5] https://www.sasb.org/, Retrieved April 11, 2020
[6] Peter Kraljic, Procurement Must Become Supply Management, Harward Business Review, Sept. 1983, https://hbr.org/1983/09/procurement-must-become-supply-management, see also (8), Retrieved April 10, 2020
[7] World Economic Forum report – Beyond Supply chain – Empowering Responsible Value Chains, published in January 2015, http://www3.weforum.org/docs/WEFUSA_BeyondSupply Chains_Report2015.pdf, Retrieved April 10, 2020
[8] Jon Lukomnik , Investor Responsibility Research Institute, State of Integrated and Sustainability Reporting 2018, 2018, https://corpgov.law.harvard.edu/2018/12/03/state-of-integrated-and-sustainability-reporting-2018/, Retrieved April 10, 2020
[9] https://integratedreporting.org, Retrieved April 10, 2020
[10] GRI and sustainability reporting, https://learn.tearfund.org/~/media/files/tilz/publications/roots/english/environmental_sustainability/roots_13_e_section_4.pdf, Retrieved April 10, 2020
[11] Gregory Unruh et al, Investing into a Sustainable Future, https://sustaincase.com/unlike-what-many-company-managers-think-for-investors-corporate-sustainability-performance-and-financial-performance-are-strongly-related-in-fact-information-on-a-companys-sustainability/, Retrieved April 10, 2020, Retrieved April 10, 2020

Chapter 12
Instead of a conclusion – a perspective?

12.1 The Paris Agreement and the 2 °C target

Climate change news already dominates the media (e.g. 1, 2) and the news published is not positive. Human activities are estimated to have caused global warming of 0.8 °C to 1.2 °C above pre-industrial levels. It is likely that global warming will reach a minimum of 1.5 °C between 2030 and 2052 if it continues to increase at the current rate.

In the Paris Agreement [7, 73], policymakers agreed on common targets to limit global warming. The agreement aims for an increase of no more than 2 °C above pre-industrial levels. (It should be noted that in autumn 2020 the European Parliament decided to aim for a target of 55% of 1990 CO_2 emissions by 2030 and in spring 2021 the German government set a target of 65% of 1990 CO_2 emissions by 2030.) The Paris Agreement has been subject of controversy: Some claim that the 2 °C target is not based on scientific evidence (e.g. [5]). Even among those who agree on the target, there is debate about how best to achieve it – what there is agreement on is that policymakers need to take drastic action to achieve it, because the reality is that CO_2 emissions are still rising globally (2020) and the likelihood of meeting emissions targets is considered to be only about 30% [6].

In other words, there is a two-thirds probability that emissions will not decline in time. In addition, the atmospheric system reacts quite slowly [4]. Taking a more realistic view of the next three decades [9, 10], that is if man-made greenhouse gas (GHG) emissions do not peak before 2030, warming of at least 3 °C must be expected.

To sketch this development a scenario:

2020–2030:
As projected by Xu and Ramanathan, carbon dioxide levels reach 437 parts per million by 2030 – unprecedented in the last 20 million years – and warming reaches 1.6 °C [8].

2030–2050:
Emissions peak in 2030 and then begin to decline, which would be associated with an 80 per cent reduction in fossil fuel energy intensity by 2100 compared to 2010.

This leads to a warming of 2.4 °C by 2050 (“baseline-fast” scenario by Xu and Ramanathan, [8]). However, due to the activation of a number of CO2 emitters, such as thawing swamps, there is a further warming of 0.6 °C bringing the total warming to 3 °C by 2050. (It should be noted that this is far from an extreme scenario: warming could (with 5% probability) exceed 3.5–4 °C by 2050 [8]).

https://doi.org/10.1515/9783110767308-012

2050:
There is now broad scientific acceptance [8] that by 2050 the tipping points for
- the West Antarctic ice sheet and a sea ice-free Arctic summer will be exceeded well before 1.5 °C warming,
- the Greenland ice sheet well before 2 °C, and
- the widespread loss of permafrost soils and the large-scale drying and dying of the Amazon by 2.5 °C.

In such a situation, the Earth is heading for warming of another degree or more, this is also because human GHG emissions will still be substantial (see the discussion of tipping points below).

By then, the world's population will perceive long-term changes (by 2100) that will be accompanied by massive migratory movements due to heat and subsequent changes in the agricultural and industrial sectors.

The analysis published consistently delivers the insight that [11, 12]:
- The 2-degree scenario will lead to a drastic change in temperature and precipitation in northern Europe and aridification in southern Europe and Africa. Aridification is also likely to occur in the USA and Central Asia.
- Food production and supply will suffer – meaning less secure and inadequate supplies and rising prices.
- Timber production will come under pressure.
- More governance arrangements can be expected – such as the EITI, which advocates for greater financial transparency and accountability of public revenues from the extractive sector, limiting the extent of corruption [14]
- Infrastructure will suffer from the heat and falling water levels. This in turn will impact on how raw materials, semi-finished and finished goods are delivered.
- The future role of digital communication is unclear, as on the one hand, energy and resource-intensive mobile phones will have to become more expensive to purchase and will be perceived as luxurious, while on the other hand, print will lose its appeal. The effect of cloud computing and mobile phone- or internet-based communication is unclear.

According to the current understanding of how democratic societies can deal with problems, it seems unlikely to solve an existential problem of this kind on a time scale fast enough to avert developments. Therefore, Anderson [12] argues for a Marshall Plan-style programme to rapidly build a CO_2-free energy supply. This would require a significant redistribution of resources from activities that support the high-carbon lifestyles of richer societies. In addition, industrial production will have to deal with the changed global scenario and there will be a number of new challenges:

- A significant proportion of the sites where production facilities are located today will be eliminated, for example due to flooding
- The supply situation deserves more attention as access to materials (i.e. infrastructure and transport) could become more difficult.
- Regulators and consumer awareness will force product lifetimes to be extended.
- The energy supply situation will become more unstable, therefore lower consumption of energy (and also other resources) needs to be used.
- The scarcity of raw materials will lead to installed industrial equipment being seen not only as a service provider but also as a resource for raw materials (see also [15]).

12.2 Tipping points

The idea of tipping points was introduced into the climate discussion by the Intergovernmental Panel on Climate Change (IPCC) more than 20 years ago [74]. Basically, these tipping points refer to individual regions which, due to special properties (e.g. their ability to store and release significant amounts of CO_2 or CH_4), can have such a large impact on the global climate in the event of a significant change in conditions that they can cause non-linear effects on a global scale. According to Lenton et al. [87], the identified tipping points and key changing properties are as follows (Table 12.1):

Table 12.1: Tipping points.

– Amazon rainforest	Frequent droughts
– Arctic sea	Ice reduction
– Atlantic circulation	Slowdown since the 1950s
– Boreal forest	Fires and pest changes
– Coral reefs	Large-scale die-off
– Greenland ice	Shelf ice loss accelerating
– Permafrost	Thawing
– West Atlantic ice sheet	Ice loss accelerating
– Wilkes Basin	Accelerating ice loss

It is not clear whether all tipping points have yet been identified and characterised in terms of key properties such as GHG storage capacity. When the concept was first introduced, it was expected that these effects would only come into play when global warming exceeded 5 °C. More recently, however, the view has changed and the IPCC now suggests that tipping points could be reached at global warming of as little as 1–2 °C. Several scientists point out that the tipping points may already have been reached [75–77].

The scenario drawn above does not leave much room for optimism. It is important to face the facts and be realistic about the likely consequences for the economy.

12.3 Economic scenario until 2050

Various aspects can be used to characterise an economy – see also Chapter 4. Here we restrict ourselves to gross domestic products (GDP), demographic development and prices.

Gross domestic product

The economic impacts of climate change and their effects on GDPs are the subject of several studies. If temperatures were to rise by only 2 °C, global domestic product would fall by 15% by 2050 compared to 2015 [16]. However, if temperatures were to rise to 3 °C, global trade would decrease by 25% [16]. It is important to note that during the Great Depression, world trade also fell by 25% – but recovered. However, the climate-induced change would be permanent, so recovery would not be expected [16]. The varying degree of economic development in different regions of the world means that emissions will continue to rise even if industrialised countries succeed in limiting CO_2 emissions.

The more industrialised the nation under consideration, the more decoupled national GDP and CO_2 emissions are [17]. In other words, the CO_2 emission curves of the more developed countries will flatten, while those of the less developed countries will continue to rise more sharply.

Different models provide different estimates of future damage from climate change and hence different impacts on the economy, ranging from 2% to about 10% or more decline in global GDP per year, depending on the increase in global average temperature. These values are based on the results of three widely used models with damage estimates based on IPPC estimated of likely temperature change by 2100 [18]. One estimate of the cost of achieving the Paris Agreement target of limiting temperature rise to no more than 2 °C assumes that about 1.5% of world income (roughly the equivalent of 1 year's real income growth) would have to be spent on reducing the global carbon footprint. However, this is under the most favourable assumptions for international cooperation. Under less favourable assumptions, the cost is estimated at over 4% of global GDP [27]. An obvious way to raise the necessary funds would be to make carbon emissions more expensive for polluters. To this end, various mechanisms have been discussed and implemented, such as taxes, (tradable) allowances as provided for by the ETS and with various support mechanisms (e.g. cap-and-trade), etc. (see Chapter 4 below). The systems have their advantages and disadvantages. The most salient aspect is that a price increase through a transparent tax should also lead to a reduction in emissions.

The impact of a CO_2 tax would, for example in the case of coal prices, with a tax burden of €44/tonne CO_2 eq., lead to coal price increases by a factor of five [27].

12.3.1 Fewer jobs and climate change lead to migration

The ILO [43] estimates that climate change threatens 1.2 billion jobs – mainly in agriculture, fisheries and forestry [19]. Because of the lack of jobs, but also because of uninhabitable areas, climate change with changes in temperature and precipitation will make some areas completely or partially uninhabitable [44, 13], which in turn will lead to migratory movements to areas that are still considered climate-stable. People will leave coastal areas, drought-stricken farmlands and areas with extreme natural disasters [44]. By 2050, climate change will force 700 million people to migrate [20, 87]. The latter source concludes that "in the next 50 years, 1 to 3 billion people will live outside the climate conditions that have served humanity well for the past 6,000 years".

The more frequent occurrence of severe weather and the uninhabitability of some areas due to overheating will require corrective measures – but it is questionable whether the necessary investment capital is available. In the private sector, the need for investment capital will cause a decline in exports. The same study estimates that prices for raw materials will rise dramatically, as the countries from which these resources originate are more affected by climate change [22]. Already, food prices are rising due to increasing heat and water shortages [23]. Crops are used as feed for animals, so meat production will be affected as well as vegetables for human consumption. In addition, productivity in related industries falls sharply, especially for outdoor work [23]. This further drives up the cost of food.

In 2017, the US Department of Defence reported [45] that defence costs are expected to rise as migration and hunger lead to social insecurity that may need to be addressed by the armed forces. In the same year, the world's 1,000 largest companies were responsible for 12% of GHG emissions. Some are starting to take action, but so far only 14% have targets that are in line with the 2 °C target, with a further 30% committing to do so in the next two years. Investment firms such as HSBC Holdings and Goldman Sachs have started to invest in more low-carbon companies [25]. For commodities, in addition to decreasing accessibility, which will lead to increased transport costs, there is the implication that climate change and possible political instability in the producing countries will limit availability and hinder the supply chain. Both factors will lead to an increase in commodity prices. This has the implication (at least) that commodity prices will rise.

12.4 Legislation and litigation

In the following, we will look at some of the laws currently in place to prevent or mitigate climate change. Some of these laws are intended for consumer goods rather than industrial goods, but can and will be extended to industrial goods once a specific legal mechanism is in place and the pressure is great enough [28]. This would also mean regulations for industrial goods such as the registration of products, the deposit of a recycling fee that will last the existence of the company in case of bankruptcy, the disclosure of sold numbers, the proper disposal of waste and so on.

Paris Agreement and translation into state legislation

In 2015, 196 countries adopted the Paris Agreement. The intention was to set the world on a course towards sustainable development. To do this, it was deemed necessary to

- change development pathways to limit warming to 1.5–2 °C above pre-industrial levels.
- increase the capacity to adapt to the adverse effects of climate change and promote climate-resilient development with low GHG emissions without jeopardising food production, and
- work towards making financial flows compatible with a pathway to low GHG emissions and climate-resilient development.

The Paris Agreement (Article 4, paragraph 2) requires each party to successively establish, communicate and maintain nationally determined contributions that it intends to achieve. Parties must pursue domestic mitigation actions to achieve the targets of these contributions. The Paris Agreement recognises that the long-term targets set out in Articles 2 and 4.1 must be achieved over time. The EU-wide targets for 2030 include [29]:

- At least 40% reduction in GHG emissions (compared to 1990 levels) – a target increased to 55% in 2020 and 2021 respectively
- At least 32% share of renewable energy
- At least 32.5% improvement in energy efficiency

The EU and other legislative bodies around the world are already taking action by enacting several laws and directives – for example, the commitment to support the use of renewable resources and to increase the share of plastic reuse to 90% by 2022 [30].

Expanding existing legal frameworks

Apart from global warming and carbon footprint reduction, it is worth noting that several other aspects are also addressed in legally binding policies or laws in the context of reducing the environmental footprint – to pick a few examples:

- Ordinance on the design of products (e.g. 43)
- Regulation forcing producers to take back products at the end of their useful life (e.g. 29)
- Prohibition of certain products
- Regulations to protect the environment for extractive industries in the exploration, extraction and handling of resources such as oil (e.g. 39)

In addition, legislation can be expected in areas that do not yet exist – for example extensions of existing legal paradigms to allow legal action to be taken when non-human – but living – creatures are affected. This was paraphrased by the headline “Should trees have standing?” in Christopher Stone’s article with the same title [39]. This concept continues to evolve – it was also applied to a lake in 2019 [40, 41].

Similarly, the rights of future generations can be seen – as these people are apparently not yet alive, they are not currently granted legal personality. On the other hand, even if they are not yet alive, this group of people could reasonably be granted a right to experience life without being exposed to massive risks knowingly imposed on them by previous generations. However, some jurisdictions have begun to recognise related rights, for example in Oposa vs. Factoran [46], the Supreme Court of the Philippines accepted that a class action can be filed on behalf of future generations to challenge logging licences. Notably, the court also affirmed that natural resources are held in trust for the benefit of present and future generations and that the government consequently has a duty to protect them [47] (see also Greta Thunberg’s speech at the United Nations; Transcript: Greta Thunberg’s speech at the UN Climate Action Summit).

12.4.1 Legal disputes and court proceedings

Climate change litigation is a new form of lawsuit. Strategic litigation aimed at holding corporations and governments accountable is becoming more common. Parties involved in such litigation increasingly include non-governmental organisations, which in turn file lawsuits against government agencies or companies or individuals on behalf of the legal entities described above [36]. For example, by linking the impacts of climate change to human rights violations, Typhoon Haiyan survivors, Greenpeace Southeast Asia and others have petitioned the Commission on Human Rights of the Philippines against the “carbon majors”.

The number of court cases dealing with environmental issues is increasing. Strategic cases aim to pressure national governments to take more ambitious climate action or enforce existing laws, while cases against large emitters seek compensation for loss and damage [31]. The United Nations Environment Programme summarised the status of climate change litigation by identifying the main “Five Trends in Climate Change Litigation Worldwide” [32] (Table 12.2):

Table 12.2: Five trends in climate change litigation worldwide (according to [32]).

1.	governments being held to their legal and political obligations
2.	linking the impacts of resource extraction to climate change
3.	proving that certain emissions are the direct cause of certain negative climate impacts
4.	establishing liability for failures (of efforts) to adapt to climate change
5.	application of the public trust doctrine to climate change

To make the required actions more tangible, an overview of the litigation already taking place has been provided in [33] an overview of the litigation already taking place (see Table 12.3):

Table 12.3: Legal measures required to address climate change impacts (according to [33]).

- compensation for the costs of adaptation to climate change
- inclusion of climate change risk in investment decisions
- disclosure of climate risk
- preventing future emissions and contributions to climate change
- requiring governments or regulators to take action to meet national or international commitments
- raising awareness and exerting pressure on corporate actors, regulators or investors

Strategic climate change lawsuits are increasingly directed against corporate actors – especially fossil fuel and cement companies, also known as "carbon majors". Underlying these lawsuits is the argument that the GHG emissions of a small group of corporations have contributed significantly to climate change over time [34, 35].

To take one example [36]:

Pacific Coast Federation of Fishermen's Associations v. Chevron Corp, filed in November 2018, seeks to hold 30 fossil fuel generation companies responsible for large losses suffered by crabbers in California and Oregon as a result of algal blooms, which in turn are attributed to global warming and poisoning shellfish. The plaintiffs demand that the defendants fund the measures necessary to sustain the crab fishing industry in the future.

With this in mind, businesses must be prepared to adapt to the changing regulatory landscape and corresponding consequences in climate change-related potential litigation. As the scope of this new category of lawsuits evolves, companies will face legal challenges related to the transition to a lower-emissions economy. The pressures created by litigation, regardless of its success or failure, must also influence the regulatory and operational environment. Case law against this backdrop may also drive changes that facilitate emissions disclosure. This, in turn, may lead to climate-related data held by governments and companies on climate change becoming publicly available and, in turn, drive climate-related lawsuits as a consequence [37].

12.5 Summarizing thoughts

Several considerations and thoughts come into mind that a reader should have concluded after studying this little booklet:

1. Economising with minimal resource consumption.
Probably there is no way to avoid a climate change more severe that was experienced so far – weather catastrophes and pandemics motivate reorientation. Subsequently, this leads to power shifts and price increases for all categories of goods. Increasing prices not only reflect this development but are a tool and will be used as such.

Today sustainability means the conscious use of resources in the face of diminishing reserves. Consideration of the natural resources available and their ecological interrelationships was and is a boundary condition of every economic activity. Economic activity with minimal resource consumption becomes a central premise of economic activity, hence.

2. Climate-neutral economy calls for a *creative* economy, not *no* economy.
Ecologically sustainable economic activity means less consumption. Awareness of sustainability is often closely tied to the fear that a climate-friendly economy is synonymous with reduction of economic activities. Given how economy operates today, less consumption requires acting and more creativity – also for alternate paths for innovation be it in the private or the industrial sector. If things are handled right less consumption can and will be associated with a gain in positive quality of life as a value add in the perception of the customer.

The current economic system will anticipate this and change – or disappear in favour of other approaches.

3. From knowledge to action – act locally.
It is a well-known fact that clear knowledge does not always lead to clear instructions for action. Moreover, interrelated and equally important challenges and the sense of urgency limit the room for error.

Usually, it is not a matter of acting on a globally correct scale: Action knowledge for climate-relevant measures can often only be developed and operationalised locally or company-specifically – this should be the focus of ecological activities in companies. Real laboratories and pilot projects support the acquisition of experiential knowledge.

4. Arrogance of the masterminds is lesser supportive – ideas must be integrated and turned to action.
It is irrelevant by whom and whether or not sustainable thinking has been advocated for a long time or only recently in the political market of opinions – in practice, it is decisive to proactively face the challenges of the necessary transformation process locally and in all consistency. Action is required in all sectors – based on but not replaced by academic considerations and discussion. Innovative new forms of cooperation between economic sectors will support the necessary transformation.

5. Approaches and technologies have a limited lifespan.
Doing business has always been a natural part of life – new innovations and phasing out the obsolete are normal and sometimes necessary. What is unusual is the, compared to the whole of human history, very rapid change needed in the supply of energy and raw materials – the demand for pace is justified by the depletion of resources.

It must be noted, however, that technologies must be socialised – that is, despite the necessary innovations, it is necessary to fall back as far as possible on technologies with which humanity is experienced. Also therefore, the knowledge of ancient and smaller cultures is worth protecting.

6. Data do not replace values or priorities, but help to understand and plan.
The implementation of the sustainable development goals of the United Nations provides the qualitative and quantitative basis both with regard to climate-relevant indicators and the probable effectiveness of measures planned for. The disaggregated footprints and their structured reduction provide the guideline for action.

It must be assumed that the supply of energy will be based almost entirely on hydropower, nuclear energy and solar energy. Political decisions currently speak against the use of nuclear energy in most countries.

7. Digital is not the new analogue.
Digital products are no substitute for analogue products. Simply transferring economic structures from the analogue to the digital space does not work – in particular the consumption of energy must be taken into account but also resource consumption needs to be taken into account.

8. Adaption is most difficult prior to fundamental change of supply systems.
With changing situations in supply and with respect to climate, problems arise suddenly and in combination. Intuitively it could be assumed that problems are greatest AFTER the changeover to new supply systems. However, experience shows that it rather is the phase PRIOR to the changeover. It is probably the case that today we are only at the beginning of a historic phase that requires such a proactive reorientation and the problems seen will increase dramatically.

9. In this context, the real problem is the steady growth of material and energy flows.
Climate change and the scarcity of resources, not only but especially fossil fuels, lead to difficult problems that cannot be thought of or solved in terms of electoral periods. The handling of these issues is accompanied by easily perceivable symptoms such as shortages and price increases while the real problem is more complex – being the steady growth of material and energy flows.

We need to change our thinking and actions to pursue effective solutions – to address short- and long-term difficulties. With no social experience to guarantee proper implementation of policies, mistakes will be made. This will lead to unrest and injustice – but a return to the conditions of the past is neither recommended nor impossible.

References

[1] e.g. www.spiegel.de/international/europe/hell-is-coming-europe-engulfed-by-massive-heat wave-a-1275268.html

[2] https://edition.cnn.com/2019/07/03/asia/india-heat-wave-survival-hnk-intl/index.html

[3] www.theguardian.com/commentisfree/2014/sep/02/limits-to-growth-was-right-new-re search-shows-were-nearing-collapse

[4] Global warming of 1.5°C, An IPCC Special Report on the impacts of global warming of 1.5 °C, October 2018 by the IPCC, Switzerland. https://www.ipcc.ch/site/assets/uploads/sites/2/2019/06/SR15_Full_Report_High_Res.pdf accessed August 25 2021

[5] A scientific critique of the two-degree climate change target, https://www.nature.com/ar ticles/ngeo2595 Nature Geoscience volume 9, pages 13–18 (2016).

[6] Unclear climate targets, Prof. Reto Knutti, 08.12.2015 (->11)

[7] https://unfccc.int/process-and-meetings/the-paris-agreement/the-paris-agreement) and follow the related path of action (https://www.ipcc.ch/sr15/

[8] Xu, Y., Ramanathan, V. 2017. "Well below 2 °C: Mitigation strategies for avoiding dangerous to catastrophic climate changes", proceedings of the National Academy of Sciences, 114 (39),10315–10323

[9] Global warming of 1.5 °C, An IPCC Special Report on the Impacts of Global Warming of 1.5 °C, October 2018 by the IPCC, Switzerland. Electronic copies of this Summary for Policymakers is available from the IPCC website www.ipcc.ch, ISBN 978-92-9169-151-7

[10] Existential Climate-related Security Risk: A Scenario Approach, Melbourne, May 2019, Spratt, D., Dunlop, I., Barrie, C., National Centre for Climate Restoration, https://breakthroughon line.org.au

[11] https://www.ethz.ch/en/news-and-events/eth-news/news/2015/12/unclear-climate-targets %20.html

[12] Anderson, K. 2019. 'Climate's holy trinity: How cogency, tenacity & courage could yet deliver on our Paris 2°C commitment', Presentation to Oxford Climate Society, 24 January 2019, accessed 18 March 2019, https://www.youtube.com/watch?v=7BZFvc-ZOa8

[13] https://advances.sciencemag.org/content/advances/3/8/e1603322.full.pdf

[14] https://www.bmz.de/rue/en/concepts_topics/GFG_Extractive_Sector/eiti/index.html

[15] IARIGAI Conference Stuttgart, Sept 2019, Designing an industrial product following criteria for circular economy. What product designers should consider - a case study based on a printing press, Michael Has, International school of paper, print media and biomaterials, 461 Rue de la Papeterie, 38402 Saint-Martin-d'Hères, France

[16] https://www.nature.com/articles/s41586-018-0071-9

[17] Raupach, M. et al, Global and regional drivers of accelerating CO2emissions, PNAS, June 2007, Vol. 104 no 24, p.: 10288ff

[18] Revesz, K.A. et al., 2014. http://www.nature.com/news/global-warming

[19] www.ilo.org/weso-greening/

[20] https://www.nytimes.com/2018/06/29/opinion/sunday/immigration-climate-change-trump.html

[21] https://www.rohma.ch/de/regulierung/rohstoffgesetz-rohg/, Schweizerisches Rohstoffgesetz ROHG, 1. April 2014;Auswirkungen der Klimaänderungen auf die Schweizer Volkswirtschaft, Schweizerisches Bundesamt für Umwelt, 2007, P 83
[22] https://www.rohma.ch/de/regulierung/rohstoffgesetz-rohg/, Schweizerisches Rohstoffgesetz ROHG, 1. April 2014;Auswirkungen der Klimaänderungen auf die Schweizer Volkswirtschaft, Schweizerisches Bundesamt für Umwelt, 2007, e.g. p 104
[23] https://www.thebalance.com/why-are-food-prices-rising-causes-of-food-price-inflation-3306099
[24] https://climateandsecurity.org/2017/11/22/defense-bill-passes-with-climate-change-and-national-security-provision/
[25] https://www.wsj.com/articles/how-companies-are-pushing-ahead-on-climate-change-targets-1510790610?shareToken=steffd42162ba04cc2985157b82bcdaea3
[26] Grilli and Yang; Pfaffenzeller; World Bank; International Monetary Fund; Organisation for Economic Cooperation and Development (OECD) statistics; Food and Agriculture Organization of the United Nations (FAO); UN Comtrade; McKinsey Global Institute analysis
[27] Nordhaus, W., quoted in The Economics of Global Climate Change, by Harris, J.M., Roach, B., Codur, A.-M., Global Development And Environment Institute Tufts University, Medford, MA 02155, http://ase.tufts.edu/gdae
[28] EU Directive WEEE (2012/19/EU)
[29] https://ec.europa.eu/clima/policies/strategies/2030_en
[30] German “Verpackungsgesetz” as per 1. January 2019, §21 (1) 2 and § 16 (Bundesanzeiger Verlag (2017): Gesetz zur Fortentwicklung der haushalts-nahen Getrennterfassung von wertstoffhaltigen Abfällen. http://www.bgbl.de/xaver/bgbl/start.xav?startbk=Bundesanzeiger_BGBl&jumpTo=bgbl117s2234.pdf
[31] Global trends in climate change litigation: 2019 snapshot, Joana Setzer and Rebecca Byrnes, Policy report; July 2019 www.lse.ac.uk/GranthamInstitute/climate-change-laws-of-the-world/
[32] The Status of Climate Change Litigation: A Global Review, United Nations Environment Programme; Columbia University, Sabin Center for Climate Change Law (2017-05); https://wedocs.unep.org/bitstream/handle/20.500.11822/20767/climate-change-litigation.pdf?sequence=1&isAllowed=y):
[33] Climate change litigation: A new class of action; https://www.whitecase.com/publications/insight/climate-change-litigation-new-class-action”
[34] Heede, R. (2014) Tracing Anthropogenic Carbon Dioxide and Methane Emissions to Fossil Fuel and Cement Producers, 1854–2010. Climatic Change, 122(1–2): 229–241).
[35] Ganguly, G., Setzer, J., Heyvaert, V. (2018) If at first you don’t succeed: Suing corporations for climate change. Oxford Journal of Legal Studies, 38: 841–868
[36] Global trends in climate change litigation: 2019 snapshot, Joana Setzer and Rebecca Byrnes, Policy report; July 2019 http://www.lse.ac.uk/GranthamInstitute/climate-change-laws-of-the-world/
[37] The Status of Climate Change Litigation: A Global Review, United Nations Environment Programme; Columbia University, Sabin Center for Climate Change Law (2017-05) https://wedocs.unep.org/bitstream/handle/20.500.11822/20767/climate-change-litigation.pdf?sequence=1&isAllowed=y
[38] Swiss Commodity Market Supervisory Authority Act, CMSAA, of 1 April 2014 or the Commodities Act, CA, of 1 April 2014; https://www.rohma.ch/en/regulation/commodities-act/
[39] Should trees have standing? Towards legal rights for natural objects (Stone, C.). https://iseethics.files.wordpress.com/2013/02/stone-christopher-d-should-trees-have-standing.pdf

[40] Erie, L., https://www.nytimes.com/2019/02/17/us/lake-erie-legal-rights.html?utm_source=Nature+Briefing&utm_campaign=b353a80080-briefing-dy-20190218&utm_medium=email&utm_term=0_c9dfd39373-b353a80080-43827413,:
[41] https://www.smithsonianmag.com/smart-news/toledo-ohio-just-granted-lake-erie-same-legal-rights-people-180971603/#Vc7oA942qs1VfaLI.99
[42] Framework Directive 2005/32/EC
[43] www.ilo.org/weso-greening/
[44] www.nytimes.com/interactive/2017/02/17/world/americas/mexico-city-sinking.html
[45] www.thebalance.com/department-of-defense-what-it-does-and-its-impact-3305982
[46] https://www.lawphil.net/judjuris/juri1993/jul1993/gr_101083_1993.html
[47] https://www.economist.com/open-future/2019/09/17/make-a-healthy-climate-a-legal-right-that-extends-to-future-generations?utm_source=Nature+Briefing&utm_campaign=9cccfcc954-briefing-dy-20190919&utm_medium=email&utm_term=0_c9dfd39373-9cccfcc954-43827413
[48] www.labelandnarrowweb.com/contents/view_breaking-news/2019-09-18/flint-cps-inks-expects-future-price-increase/- accessed 10.10.2019; 16:31 h
[49] www.britannica.com/event/Industrial-Revolution
[50] Adam Smith, An inquiry into the Nature and Causes of the Wealth of Nations, London, 1776; See Smith, Adam (1776). https://books.google.com/books?id=C5dNAAAAcAAJ&pg=PP7" /l "v=onepage&q&f=true. London: W. Strahan. 1 ed., Retrieved April 11, 2020
[51] Thomas Malthus, An Essay on The Principle of Population, https://archive.org/details/essayonprincipl00malt (1 ed.). London: J. Johnson in St Paul's Church-yard. 1798. Retrieved April 11, 2020
[52] David Ricardo, On the Principles of Political Economy and Taxation, John Murray, Albemarle-Street, London, 1817
[53] K.Marx F.Engels Manifesto of the Communist Party, London, 1848 (Marx/Engels Selected Works, Vol. One, Progress Publishers, Moscow, 1969, pp. 98-137;)
[54] Charles E. Ziegler, in https://www.sciencedirect.com/science/referenceworks/9780080970875, 2015
[55] Christien Meindertsma, Vitra Design Museum, November 2018; https://www.design-museum.de/de/ausstellungen/detailseiten/christien-meindertsma-beyond-the-surface.html Retrieved April 11, 2020
[56] Robert Bartels, The History of Marketing Thought, Columbus, Ohio, Grid, 19761965, 1976
[57] Demirdjian, Z. S., "Rise and Fall of Marketing in Mesopotamia: A Conundrum in the Cradle of Civilization," In The Future of Marketing's Past: Proceedings of the 12th Annual Conference on Historical Analysis and Research in Marketing, Leighton Neilson (ed.), CA, Longman, Association for Analysis and Research in Marketing, 2005; Clarke, J. R. in Dobbins, J. J. and Foss, P. W., The World of Pompeii, Oxford, Routledge, 2008, p. 330; Curtis, R.I., "A Personalized Floor Mosaic from Pompeii," American Journal of Archaeology, Vol. 88, No. 4, 1984, DOI: 10.2307/504744, pp. 557-566, Stable URL: https://www.jstor.org/stable/504744
[58] Bernhard London, Ending Depression by Planned Obsolescence, New York, 1932
[59] Harry S. Truman, Inauguration address; 1949; https://www.bartleby.com/124/pres53.html: Retrieved April 11, 2020
[60] Wolfgang Sachs, The archaeology of the development idea,www.burmalibrary.org/docs14/The_Archaeology_of_the_Development_Idea.pdf; Retrieved April 11, 2020
[61] Victor Papanek 1962 Design für die reale Welt. Anleitungen für eine humane Ökologie und sozialen Wandel, 1962; HYPERLINK "https://web.archive.org/web/20170409021940/http:/archiv.kultur-punkt.ch/praesentation/architektur/springer08-12papanek-design.htm"archiv.kultur-punkt.ch, Retrieved April 11, 2020

[62] Ronald L. Mace quoted in http://www.design.ncsu.edu/cud/about_ud/udhistory.htm Retrieved April 11, 2020

[63] Kenneth E. Boulding: http://arachnid.biosci.utexas.edu/courses/THOC/Readings/Boulding_SpaceshipEarth.pdf. In: Henry Jarrett (Hrsg.): Environmental Quality in a Growing Economy, Essays from the Sixth RFF Forum on Environmental Quality. Baltimore: The Johns Hopkins Press 1966. S. 3–14.

[64] David Pearce and R. Kerry Turner Economics of Natural Resources and the Environment. Johns Hopkins University Press, 1989. http://en.wikipedia.org/wiki/International_Standard_Book_Number"/o"InternationalStandardBookNumber; 978-0801839870.

[65] Walther Stahel, The Product-Life Factor, 1982; http://Product-life.org/en/major-pub-lica tions/the-product-life-factor, Retrieved April 11, 2020

[66] H. R. Sankey: The Thermal Efficiency of Steam-Engines. In: M.P.I.C.E. 125, 1896, S. 182–242.

[67] M. Has et al, Design and Management for Circularity – the Case of Paper; World Economic Forum Davos 2016; http://www.cepi.org/publication/design-and-management-circularity-%E2%80%93-case-paper, Retrieved April 11, 2020

[68] Tim Jackson, Circular Economy edited collection Clean Production Strategies, 1990; https://www.theseus.fi/bitstream/handle/10024/113985/Sudentaival_Hanne_Introduction%20to%20Circular%20Economy.pdf?sequence=1, Retrieved April 11, 2020

[69] Bhutan's Gross National Happiness Index, Oxford poverty and Human Development Initiative; https://ophi.org.uk/policy/national-policy/gross-national-happiness-index/ Retrieved April 11, 2020

[70] Schumacher, E. F.; Small Is Beautiful: Economics As If People Mattered; www.daastol.com/books/Schumacher%20(1973)%20Small%20is%20Beautiful.pdf, Retrieved April 11, 2020

[71] Schumacher, E. F.; Small Is Beautiful: Economics As If People Mattered : 25 Years Later...With Commentaries (1999). HYPERLINK "https://en.wikipedia.org/wiki/International_Standard_Book_Number"ISBN0-88179-169-5;https://en.wikipedia.org/wiki/Hartley_%26_Marks_Pub lishers"/o"Hartley&MarksPublishers

[72] Jay Forrester, World Dynamics. Wright Allen Press, Cambridge, USA, 1971, https://monoskop.org/File:Forrester_Jay_W_World_Dynamics_2nd_ed_1973.pdf, Retrieved April 11, 2020

[73] Club of Rome, Limits of Growth, 1972 (http://www.donellameadows.org/wp-content/user files/Limits-to-Growth-digital-scan-version.pdf, Retrieved April 11, 2020

[74] HYPERLINK "http://www.peakoilindia.org/wp-content/uploads/2013/10/Limits-t

[75] www.theguardian.com/commentisfree/2014/sep/02/limits-to-growth-was-right-new-re search-shows-were-nearing-collapse, Retrieved April 11, 2020

[76] Brundtlandt, G.H. "Report of the World Commission on Environment and Development: Our Common Future: From One Earth to One World" http://www.un-documents.net/wced-ocf.htm, Retrieved April 11, 2020

[77] Thomas S. Kuhn: The Structure of Scientific Revolutions, University of Chicago Press, 1970

[78] Rio Earth Summit in 1992 (Rio Convention) www.un.org/en/development/desa/population/mi gration/generalassembly/docs/globalcompact/A_CONF.151_26_Vol.I_Declaration.pdf, Retrieved April 11, 2020

[79] Eco-Design requirements of the EU, http://europa.eu/youreurope/business/product/eco-de sign/index_en.htm Retrieved April 11, 2020

[80] B. Murrey as quoted in http://www.wrap.org.uk/sites/files/wrap/Embedding%20sustainabil ity%20in%20design%20%20-%20final%20v1.pdf, Retrieved April 11, 2020

[81] The Paris Agreement, https://unfccc.int/process-and-meetings/the-paris-agreement/the-paris-agreement, Retrieved April 11, 2020

[82] Lenton, T. M. et al. Proc. Natl Acad. Sci. USA 105, 1786–1793 (2008

[83] https://www.nature.com/articles/d41586-019-03595-0, Retrieved April 11, 2020

[84] IPCC. Global Warming of 1.5°C IPCC: HYPERLINK "http://www.ipcc.ch/report/sr15/" Global Warming of 1.5 °C – an IPCC special report on the impacts of global warming of 1.5 °C above pre-industrial levels and related global greenhouse gas emission pathways, in the context of strengthening the global response to the threat of climate change, sustainable development, and efforts to eradicate poverty – SR15. In: ipcc.ch. 5. Oktober 2018, Retrieved April 11, 2020
[85] Steffen, W. et al. Proc. Natl. Acad. Sci. USA 115, 8252–8259 (2018)
[86] Rocha, J. C., Peterson, G., Bodin, Ö. & Levin, S. Science 362, 1379–1383 (2018)
[87] Lenton, T.M., Rockström, J., Gaffney, O., Richardson, K., Rahmstorf, S., Steffen, W., Schellnhuber, H.J., Climate tipping points – too risky to bet against, https://www.nature.com/articles/d41586-019-03595-0

Annex

Annex 1 – Sources

General Literature

[1] DIRECTIVE OF THE EUROPEAN PARLIAMENT AND OF THE COUNCIL ANNEX I: Method for setting generic eco-design requirements.

[2] Directive 2008/28/EC of the European Parliament and of the Council of 11 March 2008 amending Directive 2005/32/EC establishing a framework for the setting of eco-design requirements for energy-using products, as well as Council Directive 92/42/EEC and Directives 96/57/EC and 2000/55/EC, as regards the implementing powers conferred on the Commission (OJ L 81, 20.3.2008, p. 48).

[3] COM(2006) 545 final of 19 October 2006.

[4] Council Directive 92/75/EEC of 22 September 1992 on the indication by labelling and standard product information of the consumption of energy and other resources by household appliances (OJ L 297, 13.10.1992, p. 16). Directive as amended by Regulation (EC) No 1882/2003 of the European Parliament and of the Council (OJ L 284, 31. 10.2003, p. 1).

[5] Regulation (EC) No 106/2008 of the European Parliament and of the Council of 15 January 2008 on a community energy-efficiency labelling programme for office equipment (OJ L 39 of 15.1.2008, p. 1).

[6] Regulation (EC) No 1980/2000 of the European Parliament and of the Council of 17 July 2000 on a revised Community Eco-Label Award Scheme (OJ L 237 of 21.9.2000, p. 1).

[7] M. Koslowski, G. Dwalischwili, N. Marbach (Hrsg.): Eco-Design Tool. KDID, 2013, ISBN 9783000455315.
Ursula Tischner, M. Charter (Hrsg.): Sustainable Solutions. Developing Products and Services for the Future. Greenleaf, Sheffield 2001.

[8] Directive 2006/121/EC of the European Parliament and of the Council of 18 December 2006 amending Council Directive 67/548/EEC on the approximation of laws, regulations and administrative provisions relating to the classification, packaging and labelling of dangerous substances in order to adapt it to Regulation (EC) No 1907/2006 concerning the Registration, Evaluation, Authorisation and Restriction of Chemicals (REACH) and establishing a European Chemicals Agency (OJ L. L 396 of 30.12. 2006).

[9] Directive 2002/95/EC of the European Parliament and of the Council of 27 January 2003 on the restriction of the use of certain hazardous substances in electrical and electronic equipment (OJ L 37, 13.2.2003, p. 24).

[10] Council Directive 76/769/EEC of 27 July 1976 relating to restrictions on the marketing and use of certain dangerous substances and preparations (OJ L 262, 27.9.1976, p. 201).

[11] Sixth Community Environment Action Programme, laid down by Decision No 1600/2002/EC of the European Parliament and of the Council of 22 July 2002 (OJ L 242, 10.9.2002, p. 1).

[12] Communication from the Commission to the Council and the European Parliament – Integrated Product Policy – Building on Environmental Life-Cycle Thinking (IPP) (COM (2003) 302 final).

[13] The relevant rules on recasting of Community legislation apply, namely the Interinstitutional Agreement of 28 November 2001 on a more structured use of the recasting technique for legal acts, Official Journal C 077, 28/03/2002, p. 1, and Rules 80 A of the European Parliament Rules of procedure.

https://doi.org/10.1515/9783110767308-013

[14] René van Berkel, Esther Willems, and Marije Lafleur. 1997: "Development of an industrial ecology toolbox for the introduction of industrial ecology in enterprises – I" Journal of Cleaner Production 1997, 5:1–2, pp. 11–25.
[15] ISO 14 000 series will become the international standard for certifying environmental management systems. Especially life cycle assessment standard – ISO 14040-44
[16] www.biothinking.com/btintro.htm
[17] C. Vezolli, E. Manzini: Review: design for sustainable consumption and production systems. In: Arnold Tukker u. a. (Hrsg.): Perspectives on radical changes to sustainable consumption and production. (= System innovation for sustainability. 1). Greenleaf, Sheffield 2008, S. 138–158.
Karin-Simone Fuhs, Davide Brocchi, Michael Maxein, Bernd Draser (Hrsg.): Die Geschichte des Nachhaltigen Designs: Welche Haltung braucht Gestaltung? Heinrich-Böll-Stiftung / VAS – Verlag für Akademische Schriften, Bad Homburg 2013, ISBN 978-3-88864-521-1.
[18] C. Vezolli, E. Manzini: Review: design for sustainable consumption and production systems. In: Arnold Tukker u. a. (Hrsg.): Perspectives on radical changes to sustainable consumption and production. (= System innovation for sustainability. 1). Greenleaf, Sheffield 2008, S. 138–158.
Karin-Simone Fuhs, Davide Brocchi, Michael Maxein, Bernd Draser (Hrsg.): Die Geschichte des Nachhaltigen Designs: Welche Haltung braucht Gestaltung? Heinrich-Böll-Stiftung / VAS – Verlag für Akademische Schriften, Bad Homburg 2013, ISBN 978-3-88864-521-1.
[19] M. Has et al, Design and management for circularity – the case of paper; World Economic Forum Davos, 2016.
[20] http://engagebydesign.org/values/
[21] www.theRSA.org
[22] http://www.guardian.co.uk/sustainable-business/gallery/sustainable-product-design-in-pictures
[23] The Mass Balance Movement: the definitive reference for resource flows within the UK environmental economy (Biffaward – 2006). Available at www.massbalance.org
[24] Friends of the Earth Europe, Friends of the Earth Austria and SERI (2009) Overconsumption – Our use of the world's natural resources. Available at: http://www.foeeurope.org/publications/2009/Overconsumption_Sep09.pdf
[25] Assessing the environmental impacts of consumption and production: priority products and materials (UNEP 2010) ISBN 978-92-807-3084-5.
[26] Eco-Design maturity model, http://link.springer.com/chapter/10.1007/978-3-642-20183-7_35?noaccess=true
[27] http://sustainablebrands.com/news_and_views/articles/integrating-sustainable-product-design-lcapartii?utm_source=newsletter&utm_medium=innovation&utm_campaign=feb6
[28] Circular economy http://www.wrap.org.uk/content/wrap-and-circular-economy
[29] Lewis, H.; Gertsakis, J. 2001. Design + environment. A global guide to designing greener goods.
[30] Greenleaf Publishing.
[31] Tischner, U. et al. 2000. How to do eco-design? Edited by the German Federal Environmental Agency.
[32] Fuad-Luke, A. 2002. The eco-design handbook. Spanish edition edited by Cartago.
[33] Recoup (plastics recycling information) – http://www.recoup.org/business/default.asp
[34] 1IGD – sustainability information for the consumer goods industry including case studies, www.igd.com

Annex 2 – Glossary

For product-related aspects, definitions exist in the EU Directive for setting eco-design requirements (3).

Environmental footprints are quantitative measures that show the appropriation of natural resources by humans (4). Footprints are divided into ecological, economic and social footprints, as well as combined ecological, social and/or economic footprints. The idea of environmental footprints was introduced by Rees (5). An environmental footprint is divided into three aspects:

- The energy footprint – often measured in kg eq CO_2 – and therefore often referred to as the CO_2 footprint or carbon footprint. It provides a measure of the energy consumed during a defined life phase or the entire life of the product/system.
- The material footprint – It provides a measure of the material consumed during a defined life phase or the entire life of the product/system. It is often measured in kg of the material under consideration.
- The material footprint is often split into two parts: (i) which refers to material consumption (often a cascade) and (ii) which relates to hazardous waste and emissions (also often a cascade).

Life Cycle Assessment

The term "throughout the life of the product/system" is often too vague for analytical purposes. Therefore, it is useful to divide the life cycle into phases of the product's life. Strictly speaking, these phases start with the generation/extraction of the required raw materials and extend to the disposal of the materials involved. The resulting environmental footprint is the outcome of a life cycle assessment, which in turn is determined during a life cycle assessment. The EU defines Life Cycle Assessment as follows (see also below, 6, 7):

"Life Cycle Assessment (LCA) is an internationally standardised methodology. LCA helps to quantify the environmental impacts, environmental benefits, trade-offs and areas for improvement associated with goods and services (products), taking into account the whole life cycle of the product".

There are several definitions of LCA created in an attempt to explain the tool and methodology – for example:

"Compilation and assessment of the inputs, outputs and potential environmental impacts of a product system throughout its life cycle". ISO 14040:2006

"A process for evaluating the environmental aspects associated with a product over its life cycle". Goedkoop et al. (2008)

https://doi.org/10.1515/9783110767308-014

The European Commission's Eco-Design Directive and subsequent regulations (see e.g. 9) require that LCA be carried out during product design in order to minimise the environmental footprint of the product under consideration.

"Life Cycle Inventory (LCI) and Life Cycle Impact Assessment (LCIA) are sequential parts of a life cycle assessment, where

- the LCI is the collection and analysis of data on environmental impacts (e.g. emissions to e.g. air and water, waste generation and resource consumption) associated with a product from raw material extraction through production and use to disposal, including recycling, reuse and energy recovery.
- Life cycle assessment is the estimation of indicators of environmental pressures in the form of e.g. climate change, summer smog, resource use, acidification, human health impacts, etc. associated with the environmental interventions assigned to the life cycle of a product.

The data used in the LCA should be consistent, quality-assured and reflect actual industrial process chains. The methods should reflect a best consensus based on current practice.

ISO 14040:2006 defines the life cycle of a product as: "Successive and interrelated stages of a product system, from procurement of raw materials or generation from natural resources to final disposal". This "minimisation" is the result of consciously designing the product to minimise consumption – eco-design.

The term "eco-design" is often linked to the term "eco-management" in the sense that while eco-design aims to limit the addition of harmful elements to the multiple cycles of materials, eco-management limits the loss of accessible materials, which requires a consideration of the life cycle before and after the cycle.

Herman Miller, a US furniture company, describes its goals for eco-management as "Perfect Vision 2020" (10):

- Zero landfill
- Zero hazardous waste generation
- Zero air emissions (VOC)
- Zero process water consumption
- 100% green electrical energy
- 100% of turnover from DfE products
- Company buildings built to at least LEED Silver certification

An environmental audit is a management tool that involves a systematic, documented, periodic and objective assessment of the performance of the environmental organisation, management and equipment, with the aim of contributing to the protection of the environment. The objectives of the audit are to facilitate the management and control of environmental practices, and to assess compliance with company policies, which includes compliance with legal requirements.

Recycling refers to the reprocessing of waste materials in a production process for the original purpose or for other purposes, but excluding energy recovery.

"Energy recovery" means the use of combustible waste to generate energy by direct incineration with or without other waste but with recovery of the heat.

Waste means any substance or object falling within the categories listed in Annex I to Directive 2006/12/EC which the holder discards or intends or is required to discard.

Product design

Product design means the totality of the processes by which legal, technical, safety, functional, market or other requirements for a product are translated into the technical specification for that product.

Eco-design means the integration of environmental aspects into product design with the aim of improving the environmental performance of the product throughout its life cycle.

Environmental improvement

Refers to the process of improving the environmental performance of a product over successive generations, although not necessarily in relation to all environmental aspects of the product simultaneously.

Board (paperboard)

The generic term for certain types of paper, often characterised by relatively high stiffness. The primary distinction between paper and paperboard is usually based on thickness or "basis weight", although in some cases the distinction is based on properties and/or end use. For example, some lower basis weight materials, such as certain grades of folding boxboard and corrugating medium, are commonly referred to as "paperboard", while other higher-basis weight materials, such as certain grades of blotting, felting or drawing paper, are commonly referred to as "paper".

CMR

Carcinogenic, mutagenic and toxic for reproduction (in relation to substances). Construction paper product; a heavy grade of paper used for a variety of products where the paper does not return to the paper loop.

Deinking

The process of removing ink and/or toner from a printed product. This restores the optical properties of the unprinted product as best as possible.

Direct actors

Actors or designers who are in direct contact with the paper fibres.

Fibre

An elongated, tapered, thick-walled cellular unit that is the structural component of woody plants.

GHG (emissions) Greenhouse gas

Any of the atmospheric gases that contribute to the greenhouse effect by absorbing infrared radiation produced by the heating of the earth's surface by the sun. These include carbon dioxide (CO_2), methane (CH_4), nitrous oxide (N_2O) and water vapour (H_2O).

Grammage

The basis weight of paper or cardboard, which is a measure of the mass of the product per unit area and is expressed in grams per square metre.

Graphic paper

Paper produced for printing text or images.

Indirect actors

Actors or designers who are not in direct contact with the paper fibres.

Non-fibral elements

Substances such as coatings, paper fibres, salts, binders and adhesives used to achieve a specific condition of the paper fibre substrate.

Non-paper product components, accessories (e.g. add-ons).

Minerals, metals, plastics, etc. that enter the paper converting process or are combined with the final paper product.

Job trigger

The role in the paper cycle that defines products that either require the use of paper fibres because their specific use case requires them to be made from paper fibres (e.g. books, newspapers and packaging), or that require paper fibres for shipping, storage or other secondary purposes.

Packaging paper

The type of high-strength paper used for wrapping and packaging after it has been converted into packaging (cartons, bags). This includes both paper and board.

Paper

The generic term for a series of materials in the form of a continuous sheet or web, other than sheets or webs of pulp, commonly used for papermaking, pulping or nonwovens, made by applying vegetable, mineral, animal or synthetic paper fibres or mixtures thereof from a fluid suspension to a forming device, with or without the addition of other substances. Papers may be coated, impregnated or otherwise transformed during or after their manufacture without necessarily losing their identity as paper. In the conventional papermaking process, the fluid is water; however, new developments include the use of other fluids and air.

Paper product

Any product based on paper and/or board that has been printed and/or processed to fulfil its intended purpose. It may include printing inks, varnishes, lamination, any type of binding (e.g. glue, staples and threads) and non-paper material such as product samples.

PBT persistent

Bioaccumulative and toxic (as in substances or chemicals).

List of Figures

Figure 1.1 Sankey diagram for the use of paper in Europe. The figures refer to the masses in bio-kg. According to [58] —— **8**

Figure 1.2 Club of Rome Limits to Growth (LtG) – predictions in 1976 and comparison with actual values in 2014. Comparison of historical data with three Limits to Growth scenarios, for population, industrial output per capita, food per capita, global persistent pollution, fraction of non-renewable resources remaining. Based simply on the comparison of observed data and the Limits to Growth scenarios, the standard run scenario seems to better align with the observations made than the other two scenarios. It is remarkable that the model assumptions for the straightforward model (standard run) seem to allow good predictions [63] —— **12**

Figure 1.3 Carry-on-capacity of the Earth according to [68] —— **15**

Figure 3.1 Material flows in the circular economy for a technical product according to the Ellen McArthur Foundation (circular economy diagram (ellenmacarthurfoundation.org)). The different cascades refer to the most efficient form of utilisation to the ones that go along with more consumption – which may be lesser preferred. At the end of all processes, waste leads to emissions and landfill —— **28**

Figure 5.1 Circular value chain of paper in the context of printing [6]. The aim of the efforts involved in a life cycle analysis is to reduce the energy consumption and to keep the paper fibres in the value chain [7] —— **67**

Figure 5.2 Circular value chain of a printing press —— **68**

Figure 5.3 Circular value chain of the connected system of press and paper as substrate. It should be noted that a number of simplifications have already been made for this illustration and consumables are not shown in this figure —— **69**

Figure 5.4 Linking the elements in the circular value chain of the composite system of press and paper as substrate to life stages. For the sake of simplicity, the footprint of the raw materials is only partially taken into account for this discussion. The colours chosen indicate the allocation of the individual phases —— **70**

Figure 5.5 Guiding questions for the initial assessment of the likelihood that printing directly on the object is a promising idea to derive a better environmental footprint. Note that the arrow (no) of "cleaning fluid causes environmental degradation" strictly leads to a straight no if the cleaning fluid does not have an environmental clearance. The last two diamonds are not necessarily self-explanatory: The one labelled "Change of labels . . . " illustrates a case where printing directly on the object changes the packaging concept as such and thus implies a change of container. The new situation would require a new application and a comparison between the new and the old concept. The case marked "Rest of marking . . . " refers to a situation where the barrier properties of the container need to be re-evaluated due to the properties of the ink used in direct printing —— **81**

Figure 6.1 An example of using the strategy wheel to compare an existing design with a planned and actually implemented design [5] —— **102**

Figure 6.2 Direct actors in the life cycle of paper products —— **107**

Figure 6.3 Indirect actors in the life cycle of paper products —— **108**

Figure 6.4 Inputs and outputs in the life cycle of paper products —— **110**

https://doi.org/10.1515/9783110767308-015

Figure 6.5 Fibre losses (red arrows) and transfers (blue arrows) in the life cycles of paper products — **110**
Figure 6.6 Guidelines for actors in the fibres life cycle — **111**
Figure 6.7 Main environmental impacts in the life cycle of paper products (the abbreviation GHG stands for Green House Gases) — **114**
Figure 7.1 Sequence of activities for creating an organisational environmental footprint (based on [1]) — **122**
Figure 8.1 Avoiding a permanent connection to facilitate good reuse according to (based on 3) — **139**

List of Tables

Table 1.1 Technological changes during the industrial revolution according to [37] —— **1**
Table 1.2 Implementation possibilities of the fail by design concept according to [46] —— **6**
Table 1.3 Labelling guidelines for solutions as outlined by Schumacher for his small is beautiful approach, after [59] —— **9**
Table 1.4 Binding agreements of the Rio Convention —— **14**
Table 2.1 Primary activities in value chains according to [1] —— **19**
Table 2.2 Ways out of the dilemma of innovation in value chains —— **21**
Table 3.1 Different schools interpreting the circular economy in different ways (see detailed descriptions below) —— **27**
Table 3.2 Core principles of the circular economy —— **28**
Table 3.3 The ReSOLVE method as a framework that can be used as a guide on the way to a circular economy —— **35**
Table 3.4 Dimensions of the GNP approach implemented in Bhutan according to [28] —— **38**
Table 3.5 Proposed changes by the Stieglitz Commission to the GNP-based measurement of a nation's state of well-being [30] —— **39**
Table 4.1 Data to be considered in an LCA of a material product —— **45**
Table 4.2 Frequently used impact categories according to [8] —— **45**
Table 4.3 Techniques for streamlining an LCA —— **48**
Table 5.1 Scopes —— **53**
Table 5.2 CO_2 footprint of a newspaper printing company (rounded figures) producing mainly for the national market —— **54**
Table 5.3 Rounded carbon footprint of a plastic parts production facility —— **56**
Table 5.4 Risks going along with the environmental impact of consumption —— **58**
Table 5.5 Types of environmental risks as suggested by German Federal Banking Supervisory Authority (BAFIN) —— **59**
Table 5.6 Development of emissions and associated costs —— **62**
Table 5.7 Development of emissions and associated costs to be charged to the customers of the printing company under consideration, based on the data presented in Table 5.4 —— **64**
Table 5.8 Organisational and manufacturing contribution to the environmental footprint – European Directives —— **65**
Table 5.9 Generic representation of a simplified LCA – environmental footprint as energy and material calculation for the life phases of a product —— **66**
Table 5.10 Material consumption for recyclable materials (Al, Cu, Fe, certain plastics) and materials that can be reused to a large extent (according to 8) —— **71**
Table 5.11 Generic result of a life cycle assessment – environmental footprint as energy and material calculation for the lifetime of a product —— **72**
Table 5.12 Energy consumption [9, 10] —— **74**
Table 5.13 Carbon footprint usage phase (electrical energy plus paper consumption) assuming (a) user profile as above and (b) 100% use of non-recycled or recycled paper —— **75**
Table 5.14 Generic LCA result – environmental footprint as energy and material calculation for the lifetime of a product for the case of printing on paper —— **76**
Table 5.15 Main factors influencing the environmental footprint —— **77**

https://doi.org/10.1515/9783110767308-016

Table 5.16 Comparison of the CO_2 footprint assessed for inkjet with the data reported by FINAT (for production with conventional presses). Within the limits of uncertainty, the results are comparable — **77**
Table 5.17 Sensitivity analysis of waste streams as proposed by FINAT [16] — **78**
Table 5.18 Generic result of a life cycle assessment – environmental footprint as energy and material calculation for the lifetime of a product for the case of printing self-adhesive polymer labels — **79**
Table 5.19 Average material balance of cookers and ovens in % — **83**
Table 5.20 Average material balance of cookers and ovens in absolute contributions for an oven having a weight of 40 kg — **83**
Table 5.21 Total CO_2 balance for the oven — **85**
Table 5.22 Total emissions arising in connection with the publication of a digital newspaper — **89**
Table 5.23 Steps for the eco-design of a material product according to the Product Sustainability Forum, UK (see also [24]) — **92**
Table 5.24 Common first steps in eco-design — **93**
Table 6.1 Priorities for eco-design of products and processes — **98**
Table 6.2 MET Matrix to a composite insulating material according to [4,15] — **99**
Table 6.3 Criteria for the strategy wheel according to [6] — **100**
Table 6.4 Eco-design options for different steps in the life cycle of a product [2] — **102**
Table 6.5 “10 golden rules” of eco-design according to [19] — **105**
Table 6.6 Three core principles of the circular economy — **109**
Table 6.7 Types of recommendations and a transversal principle to support the circular economy — **109**
Table 6.8 Eco-design recommendations for direct actors in the life cycle of paper — **112**
Table 6.9 Eco-management recommendations for direct actors of the paper life cycle — **113**
Table 6.10 Main aspects of fibre resource, energy and waste management — **115**
Table 6.11 Order originator supply chain checklist — **115**
Table 7.1 Standards to be considered in determining the environmental footprint — **119**
Table 7.2 High-level matrix for summarising organisational and product-related aspects of the contributions to an environmental footprint — **120**
Table 7.3 Action flow for the assessment of environmental footprints — **121**
Table 7.4 Key parameters when considering organisational environmental footprint — **122**
Table 7.5 Benefits of good environmental stewardship according to [4] — **123**
Table 7.6 Established practices for reducing the environmental footprint of organisations. Corresponding efforts for office-related environmental protection should be worded in an environmental strategy — **124**
Table 7.7 Minimum content of the environmental audit results report — **127**
Table 7.8 Targets achieved to be reported in environmental audit results report — **127**
Table 8.1 Parameters to be influenced in the early product eco-design phase — **132**
Table 8.2 Analysis of material efficiency parameters and units (according to [1]) — **133**
Table 8.3 List of materials used — **136**
Table 8.4 Points to consider for implementing a material strategy in the design phase according to [3] — **138**
Table 8.5 Process before accepting a part as one for qualification of second life spare parts [3] — **139**
Table 8.6 Minimum content of an energy footprint report — **141**
Table 9.1 Interacting stakeholders in an organisation that produces material products — **143**

Table 9.2 High-level cooperation topics for manufacturing — **147**
Table 9.3 High-level cooperation topics for research and development — **151**
Table 9.4 High-level cooperation topics for facility management — **154**
Table 9.5 High-level cooperation topics for procurement — **157**
Table 9.6 Additional propositions of material products to be considered after reduction of environmental footprint — **160**
Table 9.7 High-level cooperation topics for marketing — **163**
Table 9.8 High-level design activities for quality management — **164**
Table 9.9 Targets for health and safety — **166**
Table 9.10 High-level cooperation topics for health and safety — **167**
Table 9.11 High-level cooperation activities for legal — **167**
Table 9.12 High-level cooperation topics for distribution and logistics — **168**
Table 9.13 High-level cooperation topics for field service — **169**
Table 9.14 High level cooperation topics for all stakeholders — **170**
Table 10.1 Priorities of a company's environmental performance. It should be noted that resource efficiency is therefore treated under different aspects (according to [1]) — **173**
Table 10.2 Steps for an environmental audit in an organisation [2]. It should be noted that the audit can focus on selected topics — **174**
Table 10.3 Structure of the report after the environmental assessment — **174**
Table 10.4 Benefits of environmental auditing according to [2] — **175**
Table 10.5 Questionnaire for the emission footprint according to [3] — **177**
Table 11.1 Target groups for communication on environmental activities — **181**
Table 11.2 Means to get employees on board to identify and implement solutions for a reduced environmental footprint (see also [1]) — **182**
Table 11.3 Best practice rules for sustainability communication according to McPherson [2] — **183**
Table 11.4 Minimum content of reports showing how environmental policy contributes to health and well-being without neglecting a company's economic objectives and needs — **184**
Table 11.5 Possible content of the product-related address for distribution — **185**
Table 11.6 Messages to increase reputation as a sustainable supplier — **188**
Table 11.7 Investor menu of sustainable investing strategies — **189**
Table 11.8 Benefits from responsible supply chain behaviour according to [7] — **191**
Table 11.9 Minimum content of a sustainability report — **192**
Table 11.10 Content of the environmental policy according to GRI and sustainability report [10] — **194**
Table 12.1 Tipping points — **199**
Table 12.2 Five trends in climate change litigation worldwide (according to [32]) — **203**
Table 12.3 Legal measures required to address climate change impacts (according to [33]) — **204**

Biography

After training as a skilled chemical worker, Michael Has studied physics in Regensburg and marketing at INSEAD in Fontainebleau. He received his doctorate in 1991 from the University of Regensburg in the field of biophysics with a thesis on pressure and temperature effects on hydrophobic interaction. At the FOGRA Institute in Munich, Dr Has headed the Innovation Research and Prepress departments. He founded and led the working group that later became the International Color Consortium (ICC) and was one of the founding members of the ICC. For the ICC he was their technical secretary for several years.

Dr Has served on several industrial advisory boards and participated in start-up companies. His scientific work on industry development led to numerous publications in the field of workflow management, market development in the printing industry, digital printing and colour management. For the latter, he was awarded a Mac World Award and received the Seybold Award for Innovation with the ICC.

After joining Océ in 1998, he held senior positions in research and development as well as product strategy, partner management, marketing and product line management. He led several new product/portfolio developments and successful market entries in software and hardware.

In parallel to his activities in industry, Dr Has habilitated at the Institut National Polytechnique of the University of Grenoble and finalised his 'Habilitation à Diriger des Recherches' in 1998.

Since 1998 he has been teaching in Grenoble as Distinguished Professor. Teaching topics include new technologies and business and portfolio strategy. Starting with the analysis of the circular economy of paper, he has been working on the focus of sustainability since around 2011. In this context, he deals with the creation of environmental footprints (CO_2 and material footprints) for companies and individual products, measures to reduce footprints and eco-design for products. In doing so, he worked as a lecturer with companies and products from very different sectors such as the automotive, chemical or printing industries.

In recognition of his contribution and his way of guiding students, he was appointed as a trusted lecturer of the Hans Böckler Foundation – the foundation of the German Trade Union Confederation. Dr Has heads the Foundation Diversity of Cultures of the World (Atiftung Vielfalt der Kulturen der Welt) – a foundation dealing with and supporting human rights activities and projects for ethnic and religious minorities.

(See also: http://pagora.grenoble-inp.fr/fr/annuaire/michael-has, https://de.wikipedia.org/wiki/Michael_Has)

https://doi.org/10.1515/9783110767308-017

Index

Adam Smith 2, 4–5, 7
alienation 4–5
Amory Lovins 27
Arnold Toynbee 1
audits 141, 173, 176

backing oven 82
BAFIN 57, 59, 65
best practices 124
biomimicry 27, 30
blue economy 27, 30–31
Brundtland 11, 13
Brundtland Commission 11, 13
business model for sustainability 25
business models 25

carbon footprint 20, 51–54, 56–57, 63, 65, 70, 73–74, 78–79, 81, 88, 139, 150, 165, 190, 200, 202, 217
carbon intensity value 73
carry-on-capacity 14
CDP 20
central planning 4–5
Christopher Stone 203
circular economy V, 6–7, 8, 14, 19, 22–23, 26–29, 34, 38, 43, 97, 106–109, 114–115, 137, 143, 229
circular value chain 21
climate change 39, 52, 60, 123, 163, 165, 200–205, 218
Club of Rome 3, 11–12, 43
CO_2 footprint 26, 51, 54, 61, 77, 84, 217
code of conduct 156, 159, 184–185
Commission on the Measurement of Economic Performance and Social Progress 38
Communication 163, 182–185, 187–188, 190
Conrad Luttropp 104
consumables 74, 76, 79, 90, 133
cooperation topics 147, 151, 154, 157, 163
corporate sustainability 32, 182, 192
cradle to cradle 163
cradle-to-cradle 22, 27
cradle-to-grave 1, 21–22

Daniel Halbheer 159
David Pearce 7
David Ricardo 3
Declarable Substances List 145, 149
deinkability 74, 116
Dennis Meadows 10
digital newspaper 86
direct labelling 80
disassembling 22
dismantling 85, 137–138, 152

Earth as a spaceship 27
eco-design V, 13, 47, 81, 92–93, 97–99, 101, 104–106, 108–109, 132, 136, 217–218, 229
eco-design options 102
eco-design recommendations 109, 111–112
ecological footprint 66
eco-management 109, 111, 113
Ellen McArthur Foundation 30
EMAS 173
emission pricing 34
emissions and associated costs 62, 64
energy and material strategy 131
energy footprint report 141
environmental assessment 174
environmental audits 125
environmental footprint V, 20–21, 65–66, 72, 75–76, 78–82, 109, 119–124, 131–132, 144–145, 156, 159–162, 168–170, 176, 178, 182, 186–188, 202, 217–218
environmental strategy 181
Ernst Schumacher 8
EU Directive 97, 217

fail by design 6
FAO 173, 175
fibres 67, 74, 106–107, 109, 114, 220–221
FINAT 75, 77–79
footprint strategy 179

G. Pauli 27
GHG 26, 44, 51–52, 60–61, 86, 92, 114, 119, 140, 176–179, 199, 220

https://doi.org/10.1515/9783110767308-018

GHGs 51
GHP 51–52, 81
Global Reporting Initiative 20, 184
GNH 37–38
GNP 7–8, 37
Great Depression 6, 200
Green Logistics 165
Green Procurement 150
Green Procurement Standards 146, 173, 186
Greenhouse Gas Protocol 51–52, 54, 65, 81
greenhouse gases 51, 119

Hans Carl von Carlowitz 2
Harold Smith 43
Harry S. Truman 6
Health and Safety 148, 152, 158, 162–164, 166–168

ICC 173–174, 229
impact categories 44–45, 48, 121
industrial ecology 27
industrial metabolism 27
industrial revolution 1, 4
IPPC 200
ISO 43, 47, 52, 59, 116, 119, 131, 166, 184, 217–218

James Watts 1
Janine Benyus 27
Jay Forrester 10
Joseph Stiglitz 38

K. Boulding 7
K. E. Boulding 27
Kerry Turner 7
King of Bhutan 37–38
Kyoto Protocol 14, 140

labels 71, 75, 78–80, 82, 159–161, 188
labels 81
LCA 14, 43–45, 47–48, 65–66, 76, 80, 85, 98, 121, 170, 217–218
LCA 43
LCIA 44, 47–48, 218
liability 164, 167, 176, 190, 203
life cycle assessment 14, 43, 48, 65, 88, 217
life cycle impact assessment 44, 218
Limits to Growth 11, 43
Ludwig v. Mises 5

Make-To-Assemble 145
Make-To-Order 145
Make-To-Stock 145
Marx 3–5
Meindertsma 5
MET Matrix 98–99
Michael Braungart 27
Michael Porter 19
migration 201
mitigation 52, 174–175, 202

National Sacrifice Areas 3
natural capital 27–31, 109
natural capitalism 27, 29–30
newspaper 53–54, 61–62, 86–89, 91, 93

Organisation Eco-footprint 119
Organisational Footprint 65

P. Laybourn 27
Paris Agreement 14, 197, 200, 202
Paris Climate Agreement 60
performance economy 27
permaculture 32
physical risks 58
planned obsolescence 6
print 66
procurement 93, 120, 143, 146–159, 164, 167–168, 170–171, 173, 186, 194
Product Sustainability Forum 92

quality management 159

REACH 146, 152, 167
read 88
recycled paper 53, 71, 74–77, 107, 116
regulatory framework 62
reputational risks 58
ReSOLVE methodology 34
Restricted Substance List 148–149, 152, 158, 164, 167–168, 170
return logistics 131
Rio Convention 13–14
risk 2, 20, 54, 58–60, 62–63, 65, 116, 149, 156, 160, 162, 189, 191, 204
RoHS 146, 152, 167
Ronald L. Mace 8

safety data sheets 162
Sankey 7–8
Sankey diagram 7–8
scope 1, 2 and 3 51
scope 1 51–54, 56, 62, 65, 89–90, 120
scope 2 53–56, 62, 89, 120
scope 3 53–56, 62, 89, 120
SDGs 32–33, 184
server farms 86, 88, 91
service 99, 120, 133, 147–148, 151–153, 158, 163, 166–167, 169
Small Is Beautiful 8
Spaceship Earth 7
Stiglitz Commission 38
strategy wheel 92, 99–102
supply chain 19–20, 21, 44, 47, 58, 62–63, 65, 91, 94, 98, 104, 106, 115, 132, 136, 141, 143, 145–146, 162, 164–165, 167–168, 171, 178, 186, 188, 190–191, 201
sustainability-related risks 59
sustainable development goals 32, 206

Thomas Grädel 27
Thomas Malthus 3
Tim Jackson 8
tipping points 198–199
transition risks 58
Truman Doctrine 6, 8, 10, 13

UN Gap Report 60
UNEP 173

value chain 7, 19–21, 25, 28–29, 52–53, 61–63, 66–70, 75, 105–106, 108–109, 111, 116, 143, 149, 163
value cycle 19
VDI Guideline 2243 34
Victor Papanek 8

Walter Stahel 27
Walther Stahel 7
William McDonough 27
World Economic Forum 105